Active Materials for Medical Applications

Active Materials for Medical Applications

Editors

Nicanor Cimpoesu
Ramona Cimpoesu

MDPI • Basel • Beijing • Wuhan • Barcelona • Belgrade • Manchester • Tokyo • Cluj • Tianjin

Editors
Nicanor Cimpoesu
Technical University
Gheorghe Asachi din Iasi
Romania

Ramona Cimpoesu
Technical University
Gheorghe Asachi din Iasi
Romania

Editorial Office
MDPI
St. Alban-Anlage 66
4052 Basel, Switzerland

This is a reprint of articles from the Special Issue published online in the open access journal *Applied Sciences* (ISSN 2076-3417) (available at: https://www.mdpi.com/journal/applsci/special_issues/medical_active_materials).

For citation purposes, cite each article independently as indicated on the article page online and as indicated below:

LastName, A.A.; LastName, B.B.; LastName, C.C. Article Title. *Journal Name* **Year**, *Volume Number*, Page Range.

ISBN 978-3-0365-5267-5 (Hbk)
ISBN 978-3-0365-5268-2 (PDF)

© 2022 by the authors. Articles in this book are Open Access and distributed under the Creative Commons Attribution (CC BY) license, which allows users to download, copy and build upon published articles, as long as the author and publisher are properly credited, which ensures maximum dissemination and a wider impact of our publications.

The book as a whole is distributed by MDPI under the terms and conditions of the Creative Commons license CC BY-NC-ND.

Contents

Nicanor Cimpoeșu and Ramona Cimpoeșu
Special Issue on "Active Materials for Medical Applications"
Reprinted from: *Appl. Sci.* **2022**, *12*, 8440, doi:10.3390/app12178440 . . . . **1**

Ramona Cimpoeșu, Petrică Vizureanu, Ioan Știrbu, Alina Sodor, Georgeta Zegan, Marius Prelipceanu, Nicanor Cimpoeșu and Nicoleta Ioanid
Corrosion-Resistance Analysis of HA Layer Deposited through Electrophoresis on Ti4Al4Zr Metallic Substrate
Reprinted from: *Appl. Sci.* **2021**, *11*, 4198, doi:10.3390/app11094198 . . . . **5**

Sung Sik Nam, Jeong Su Kim and Sang Don Mun
Magnetic Abrasive Finishing of Beta-Titanium Wire Using Multiple Transfer Movement Method
Reprinted from: *Appl. Sci.* **2020**, *10*, 6729, doi:10.3390/app10196729 . . . . **15**

Cătălin Panaghie, Ramona Cimpoeșu, Georgeta Zegan, Ana-Maria Roman, Mircea Catalin Ivanescu, Andra Adorata Aelenei, Marcelin Benchea, Nicanor Cimpoeșu and Nicoleta Ioanid
In Vitro Corrosion Behavior of Zn3Mg0.7Y Biodegradable Alloy in Simulated Body Fluid (SBF)
Reprinted from: *Appl. Sci.* **2022**, *12*, 2727, doi:10.3390/app12052727 . . . . **29**

Bogdan Istrate, Corneliu Munteanu, Ramona Cimpoesu, Nicanor Cimpoesu, Oana Diana Popescu and Maria Daniela Vlad
Microstructural, Electrochemical and In Vitro Analysis of Mg-0.5Ca-xGd Biodegradable Alloys
Reprinted from: *Appl. Sci.* **2021**, *11*, 981, doi:10.3390/app11030981 . . . . **49**

Alexandru Cocean, Iuliana Cocean, Nicanor Cimpoesu, Georgiana Cocean, Ramona Cimpoesu, Cristina Postolachi, Vasilica Popescu and Silviu Gurlui
Laser Induced Method to Produce Curcuminoid-Silanol Thin Films for Transdermal Patches Using Irradiation of Turmeric Target
Reprinted from: *Appl. Sci.* **2021**, *11*, 4030, doi:10.3390/app11094030 . . . . **67**

Koji Kubota, Akie Kakishita, Mana Okasaka, Yuka Tokunaga and Sadaki Takata
Effect of Alkyl Structure (Straight Chain/Branched Chain/Unsaturation) of C18 Fatty Acid Sodium Soap on Skin Barrier Function
Reprinted from: *Appl. Sci.* **2020**, *10*, 4310, doi:10.3390/app10124310 . . . . **79**

Yehree Kim, Jeon Min Kang, Ho-Young Song, Woo Seok Kang, Jung-Hoon Park and Jong Woo Chung
Self-Expandable Retainer for Endoscopic Visualization in the External Auditory Canal: Proof of Concept in Human Cadavers
Reprinted from: *Appl. Sci.* **2020**, *10*, 1877, doi:10.3390/app10051877 . . . . **91**

Elena-Raluca Baciu, Ramona Cimpoeșu, Anca Vițalariu, Constantin Baciu, Nicanor Cimpoeșu, Alina Sodor, Georgeta Zegan and Alice Murariu
Surface Analysis of 3D (SLM) Co–Cr–W Dental Metallic Materials
Reprinted from: *Appl. Sci.* **2021**, *11*, 255, doi:10.3390/app11010255 . . . . **99**

Ioana-Andreea Sioustis, Mihai Axinte, Marius Prelipceanu, Alexandra Martu, Diana-Cristala Kappenberg-Nitescu, Silvia Teslaru, Ionut Luchian, Sorina Mihaela Solomon, Nicanor Cimpoesu and Silvia Martu
Finite Element Analysis of Mandibular Anterior Teeth with Healthy, but Reduced Periodontium
Reprinted from: *Appl. Sci.* **2021**, *11*, 3824, doi:10.3390/app11093824 . . . . **115**

Monica Tatarciuc, George Alexandru Maftei, Anca Vitalariu, Ionut Luchian, Ioana Martu and Diana Diaconu-Popa
Inlay-Retained Dental Bridges—A Finite Element Analysis
Reprinted from: *Appl. Sci.* **2021**, *11*, 3770, doi:10.3390/app11093770 . . . . **133**

Editorial

Special Issue on "Active Materials for Medical Applications"

Nicanor Cimpoeșu [1,2,*] and Ramona Cimpoeșu [1]

1. Faculty of Material Science and Engineering, Materials Science Department, "Gheorghe Asachi" Technical University of Iasi, 59A Mangeron Boulevard, 700050 Iasi, Romania
2. Faculty of Physics, Atmosphere Optics, Spectroscopy and Laser Laboratory (LOASL), "Alexandru Ioan Cuza" University of Iasi, 11 Carol I Boulevard, 700506 Iasi, Romania
* Correspondence: nicanor.cimpoesu@tuiasi.ro

Citation: Cimpoeșu, N.; Cimpoeșu, R. Special Issue on "Active Materials for Medical Applications". *Appl. Sci.* **2022**, *12*, 8440. https://doi.org/10.3390/app12178440

Received: 5 August 2022
Accepted: 22 August 2022
Published: 24 August 2022

Publisher's Note: MDPI stays neutral with regard to jurisdictional claims in published maps and institutional affiliations.

Copyright: © 2022 by the authors. Licensee MDPI, Basel, Switzerland. This article is an open access article distributed under the terms and conditions of the Creative Commons Attribution (CC BY) license (https://creativecommons.org/licenses/by/4.0/).

This Special Issue was proposed by engineers, physicists, medical doctors, researchers and scientists. We intend to analyze and discuss different topics on active materials for medical applications. There is great potential in the application of active or smart materials (metallic, polymer or ceramic) for the progression of applications in the medical domain of MEMS, actuators, sensors or functional systems. Active or "smart" materials have the ability to respond to different physical or chemical stimuli in a specific, repeatable mode. The actual activity in the domain, however, presents problems connected to obtaining and processing, characterizing, modeling and simulating or prototyping technologies. This Special Issue of *Applied Sciences* focuses on the most recent advances in obtaining and thermal and mechanical processing active materials used in the medical field with enhanced performances.

In this volume, two papers deal with biodegradable metallic materials based on Mg and Zn (Microstructural, Electrochemical and In Vitro Analysis of $Mg_{0.5}Ca_xGd$ Biodegradable Alloys [1] and In Vitro Corrosion Behavior of $Zn_3Mg_{0.7}Y$ Biodegradable Alloy in Simulated Body Fluid (SBF) [2]), presenting findings in this new important field with new chemical systems, degradation rate determination and main physical, chemical and mechanical properties.

Biodegradable metallic materials represent a new class of biocompatible materials for medical applications based on numerous advantages. Among them, those based on zinc have a rate of degradation close to the healing period required by many clinical problems, which makes them more suitable than those based on magnesium or iron. The poor mechanical properties of Zn can be significantly improved by the addition of Mg and Y. In this research study, we analyze the electro-chemical and mechanical behavior of a new alloy based on $Zn_3Mg_{0.7}Y$ compared with pure Zn and Zn_3Mg materials [2].

Microstructure and chemical compositions were investigated by electron microscopy and energy dispersive spectroscopy. Electrochemical corrosion was analyzed by linear polarization (LP), cyclic polarization (CP) and electrochemical impedance spectroscopy (EIS). For hardness and scratch resistance, a microhardness tester and a scratch module were used. Findings revealed that the mechanical properties of Zn improved with the addition of Mg and Y. Zn, Zn-Mg and Zn-Mg-Y alloys in this study showed highly active behaviors in SBF with uniform corrosion [2]. Zinc metals and their alloys with magnesium and yttrium showed a moderate degradation rate and can be considered as promising biodegradable materials for orthopedic application.

Two other papers are based on the Finite element method (FEM) of investigation with exceptional findings in the field of mandibular anterior teeth with healthy but reduced periodontium and inlay-retained dental bridges [3,4].

Finite element analysis studies have been of interest in the field of orthodontics, and this is due to the ability to study stress in the bone, periodontal ligament (PDL), teeth and the displacement in the bone by using this method. Our study aimed to present a method that determines the effect of applying orthodontic forces in a bodily direction on a healthy and reduced periodontium and to demonstrate the utility of finite element analysis. Using the cone-beam computed tomography (CBCT) of a patient with a healthy and reduced periodontium, we modeled the geometric construction of the contour of elements necessary for the study [4].

Afterwards, we applied a force of 1 N and a force of 0.8 N in order to achieve bodily movements and to analyze the stress in the bone, in the periodontal ligament and in absolute displacement. The analysis of the applied forces showed that a minimal ligament thickness is correlated with the highest value of the maximum stress in the PDL and decreased displacement [4]. This confirms the results obtained in previous clinical practice, confirming the validity of the simulation. During orthodontic tooth movements, the morphology of the teeth and of the periodontium should be taken into account. The effect of orthodontic forces on a particular anatomy can be studied using FEA, a method that provides real data. This is necessary for proper treatment planning and its particularization depends on the patient's particular situation [4].

Other interesting topics of the special materials used in medical field cover the following: Surface Analysis of 3D (SLM) Co–Cr–W Dental Metallic Materials [5], Laser Induced Method to Produce Curcuminoid-Silanol Thin Films for Transdermal Patches Using Irradiation of Turmeric Target [6], Corrosion-Resistance Analysis of HA Layer Deposited through Electrophoresis on Ti4Al4Zr Metallic Substrate [7], Magnetic Abrasive Finishing of Beta-Titanium Wire Using Multiple Transfer Movement Method [8], Effect of Alkyl Structure (Straight Chain/Branched Chain/Unsaturation) of C18 Fatty Acid Sodium Soap on Skin Barrier Function [9] or Self-Expandable Retainer for Endoscopic Visualization in the External Auditory Canal: Proof of Concept in Human Cadavers [10].

A new possible method to produce a transdermal patch is proposed [6]. The study refers to the pulsed laser deposition method (PLD) applied on turmeric target in order to obtain thin layers. Under high-power laser irradiation of 532 nm wavelength, thin films containing curcuminoids were obtained on different substrates such as glass and quartz (laboratory investigation) and hemp fabric (practical application). Compared with FTIR, SEM-EDS and LIF analyses proved that the obtained thin-film chemical composition mainly comprises demethoxy curcumin and bisdemethoxycurcumin, which is evidence that most of the curcumin from turmeric has been demethixylated during laser ablation. Silanol groups with known roles in dermal reconstruction are evidenced in both turmeric target and curcuminoid thin films. UV–VIS reflection spectra show the same characteristics for all curcuminoid thin films, indicating that the method is reproducible [6].

The method proves to be successful for producing a composite material, namely curcuminoid transdermal patch with silanol groups, using directly turmeric as target in the thin film deposited by pulsed laser techniques. Double-layered patch curcuminoid–silver was produced under this study, proving compatibility between the two deposited layers. The silver layer added on the curcuminoid-silanol layer aimed to increase antiseptic properties relative to the transdermal patch [6].

Titanium is often used in various important applications in transportation and the healthcare industry. The goal of this study was to determine the optimum processing of magnetic abrasives in beta-titanium wire, which is often used in frames for eyeglasses because of its excellent elasticity among titanium alloys [8].

To check the performance of the magnetic abrasive finishing process, the surface roughness (Ra) was measured when the specimen was machined at various rotational speeds (700, 1500 and 2000 rpm) in the presence of diamond pastes of various particle sizes (0.5, 1 and 3 µm). We concluded that the surface roughness (Ra) was the best at 2000 rpm, 1 µm particle size and 300 s processing time, and the surface roughness of β-titanium improved from 0.32 to 0.05 µm. In addition, optimal conditions were used to test the influence of the finishing gap, and it was found that the processing power was superior at a gap of 3 mm than at 5 mm when processing was conducted for 300 s [8].

A study was conducted to investigate the efficacy of a self-expandable retainer (SER) for the endoscopic visualization of the external auditory canal (EAC). Tympanomeatal flap

(TMF) elevation was performed in six cadaveric heads. Two different types of SER were placed. Procedural feasibility was assessed by using endoscopic images. Technical success rate, procedure time, endoscopy lens cleaning, and the presence of mucosal injuries were analyzed. TMF elevation and SER placement were successful in all specimens, and there were no procedure-related complications.

The mean procedure time with the SERs was significantly shorter than without ($p < 0.001$). The mean number of times at which the endoscopy lens was cleaned during the procedure was significantly lower in the SER group ($p < 0.001$). In the SER group, endoscopy insertion into the EAC was easier without tissue contact with the lens during the TMF elevation compared with the non-SER group. There were no mucosal injuries. SER placement is effective for endoscopic visualization via the expanded and straightened EAC. A fully covered type of SER is preferable. The device can be useful for endoscopic ear surgery, reducing procedure time and reducing the need for endoscopy lens cleaning during the procedure [8].

Author Contributions: Conceptualization, N.C. and R.C., writing—review and editing, N.C. and R.C.; visualization, N.C. and R.C.; supervision, N.C. and R.C., project administration, N.C. and R.C. All authors have read and agreed to the published version of the manuscript.

Funding: This research was funded partially by Ministry of Research, Innovation and Digitization, CNCS-UEFISCDI grant number PN-III-P1-1.1-TE-2021-0702 and by project FAIR_09/24.11.2020 and ROBIM-PN-III-P4-ID-PCE2020-0332.

Acknowledgments: Special thanks to all authors and all peer reviewers involved in the publication process for their valuable contributions to this Special Issue 'Active Materials for Medical Applications'.

Conflicts of Interest: The authors declare no conflict of interest.

References

1. Istrate, B.; Munteanu, C.; Cimpoesu, R.; Cimpoesu, N.; Popescu, O.-D.; Vlad, M.D. Microstructural, Electrochemical and In Vitro Analysis of Mg-0.5Ca-xGd Biodegradable Alloys. *Appl. Sci.* **2021**, *11*, 981. [CrossRef]
2. Panaghie, C.; Cimpoeșu, R.; Zegan, G.; Roman, A.-M.; Ivanescu, M.-C.; Aelenei, A.-A.; Benchea, M.; Cimpoeșu, N.; Ioanid, N. In Vitro Corrosion Behavior of Zn3Mg0.7Y Biodegradable Alloy in Simulated Body Fluid (SBF). *Appl. Sci.* **2022**, *12*, 2727. [CrossRef]
3. Tatarciuc, M.; Maftei, G.A.; Vitalariu, A.; Luchian, I.; Martu, I.; Diaconu-Popa, D. Inlay-Retained Dental Bridges—A Finite Element Analysis. *Appl. Sci.* **2021**, *11*, 3770. [CrossRef]
4. Sioustis, I.-A.; Axinte, M.; Prelipceanu, M.; Martu, A.; Kappenberg-Nitescu, D.C.; Teslaru, S.; Luchian, I.; Solomon, S.M.; Cimpoesu, N.; Martu, S. Finite Element Analysis of Mandibular Anterior Teeth with Healthy, but Reduced Periodontium. *Appl. Sci.* **2021**, *11*, 3824. [CrossRef]
5. Baciu, E.-R.; Cimpoeșu, R.; Vițalariu, A.; Baciu, C.; Cimpoeșu, N.; Sodor, A.; Zegan, G.; Murariu, A. Surface Analysis of 3D (SLM) Co–Cr–W Dental Metallic Materials. *Appl. Sci.* **2021**, *11*, 255. [CrossRef]
6. Cocean, A.; Cocean, I.; Cimpoesu, N.; Cocean, G.; Cimpoesu, R.; Postolachi, C.; Popescu, V.; Gurlui, S.O. Laser Induced Method to Produce Curcuminoid-Silanol Thin Films for Transdermal Patches Using Irradiation of Turmeric Target. *Appl. Sci.* **2021**, *11*, 4030. [CrossRef]
7. Cimpoeșu, R.; Vizureanu, P.; Știrbu, I.; Sodor, A.; Zegan, G.; Prelipceanu, M.; Cimpoeșu, N.; Ioanid, N. Corrosion-Resistance Analysis of HA Layer Deposited through Electrophoresis on Ti$_4$Al$_4$Zr Metallic Substrate. *Appl. Sci.* **2021**, *11*, 4198. [CrossRef]
8. Nam, S.S.; Kim, J.S.; Mun, S.D. Magnetic Abrasive Finishing of Beta-Titanium Wire Using Multiple Transfer Movement Method. *Appl. Sci.* **2020**, *10*, 6729. [CrossRef]
9. Kubota, K.; Kakishita, A.; Okasaka, M.; Tokunaga, Y.; Takata, S. Effect of Alkyl Structure (Straight Chain/Branched Chain/Unsaturation) of C18 Fatty Acid Sodium Soap on Skin Barrier Function. *Appl. Sci.* **2020**, *10*, 4310. [CrossRef]
10. Kim, Y.; Kang, J.M.; Song, H.-Y.; Kang, W.S.; Park, J.H.; Chung, J.W. Self-Expandable Retainer for Endoscopic Visualization in the External Auditory Canal: Proof of Concept in Human Cadavers. *Appl. Sci.* **2020**, *10*, 1877. [CrossRef]

Article

Corrosion-Resistance Analysis of HA Layer Deposited through Electrophoresis on Ti4Al4Zr Metallic Substrate

Ramona Cimpoeșu [1], Petrică Vizureanu [1], Ioan Știrbu [1], Alina Sodor [2,*], Georgeta Zegan [2], Marius Prelipceanu [3,*], Nicanor Cimpoeșu [1,*] and Nicoleta Ioanid [2]

1. Faculty of Materials Science and Engineering, "Gh. Asachi" Technical University from Iasi, 700050 Iași, Romania; ramona.cimpoesu@tuiasi.ro (R.C.); peviz2002@yahoo.com (P.V.); ionutstirbu@yahoo.com (I.Ș.)
2. Faculty of Dental Medicine, "Grigore T. Popa" University of Medicine and Pharmacy, 700115 Iasi, Romania; georgetazegan@yahoo.com (G.Z.); nicole_ioanid@yahoo.com (N.I.)
3. Integrated Center for Research, Development and Innovation in Advanced Materials, Nanotechnologies, and Distributed Systems for Fabrication and Control, Department of Computers, Electronics and Automation, Ștefan cel Mare University of Suceava, 720225 Suceava, Romania
* Correspondence: alinasodor@yahoo.com (A.S.); marius.prelipceanu@usm.ro (M.P.); nicanor.cimpoesu@tuiasi.ro (N.C.)

Abstract: An alloy surface with possible applications in the medical field, Ti4A14Zr, was improved through the deposition of a thin hydroxyapatite (HA) layer. In this paper, we analyzed the growth of a HA layer through electrophoresis and the corrosion resistance of the metallic sample covered with the ceramic layer. The substrate surface was processed via chemical procedures before the HA deposition. The state of the metallic surface and that of the layer of HA were investigated using scanning electron microscopy (SEM) and energy dispersive spectroscopy (EDS) analysis of the chemical composition. The results indicate a high increase in the corrosion resistance associated with the ceramic layer compared to the metallic basic layer. Moreover, the analysis revealed the formation of a homogeneous TiO_2 layer on the surface of the metallic substrate. The titanium oxide layer identified by SEM–EDS and confirmed by EIS was very homogeneous and resistant, with a compact microstructural appearance and submicron dimension. The layer composed of TiO_2 and HA provided good corrosion protection.

Keywords: Ti4Al4Zr; HA; electrophoresis; corrosion

1. Introduction

Titanium (Ti) and Ti-based alloys generally present desirable mechanical characteristics, corrosion resistance, and biocompatibility [1–3]. Titanium-based alloys have been extensively used in the last four decades to create implant elements and medical tools like prostheses or dental implants. The corrosion resistance and, at a certain level, antibacterial properties of titanium-based alloys in biomedical applications are mainly due to the formation of a passive TiO_2 layer on the surface [4,5]. The domain of oral implants involves many medical fields, with breakthroughs made in surgery, prosthetics, and dental prosthetics, in particular, offering new possibilities in the restoration of prosthetics [6,7]. There are few published experimental results about the chemical, electrochemical, terminal, or combined features of surface preparation of Ti-based alloys that facilitate both the desired chemical, mechanical, and biological reactions at a superficial level and the deposition of nonmetallic layers [8–12]. TiAlZr systems have gained some attention as implantable materials in recent years and are considered a potential solution for medical applications where Ti is not applicable [3–5,13,14]. Hydroxyapatite (HA) in large or small quantities is not toxic to biological cells) and is almost inactive in biological environments. Additionally, numerous studies have investigated and confirmed the excellent properties of hydroxyapatite in contact with bone tissues [15,16]. The most efficient method of enhancing the osteogenesis of Ti-based alloys in medical implants is covering them with a bioactive layer such as hydroxyapatite. Electrophoresis deposition is a technique that uses the motion of charged particles in a suspension in the presence of an electric field. This electric field allows the formation of well-established particles in thin layers, modified in layers regardless of their shape, or even in thicker layers on their own. The first applications of electrophoresis were in the modeling of ceramic materials and in the production of coatings. There is also a growing interest in the use of this technique in the handling of biomaterials and biological components such as natural polymers, proteins, bacteria, and cells. The process is useful for the application of polymeric materials, pigments, dyes, ceramics, or metals on any surface of an electrically conductive material. The microstructural, chemical, and mechanical characteristics of complex elements obtained via superficial layer deposition remain to be determined [17–19].

The paper deals with a chemical treatment method for the surface of an implantable metallic material based on Ti, deposition of a thin layer of hydroxyapatite (HA) through electrophoresis, and analysis of the resistance to electrocorrosion of the resulting films using EIS. Surface investigations were performed via 2D and 3D SEM.

2. Materials and Methods

The surface roughness of two cylindrical samples of Ti4Al4Zr, which were acquired from Zirom S.R:L. [20], was modified using a chemical solution of H_2SO_4 (34%) + HCl (14%) at a temperature of 65 °C for 90 min and 60 min for samples named D1 and D2, respectively. The process of engraving with acid followed by chemical activation with NaOH was performed in a thermo-stated glass tube cell. The thermo-stating was executed in a block of thermo-stated tubes of type BlockTherm MTA Kutesz 660, Hungary, which allowed the temperature to be maintained with a ± 1 °C precision.

For the growth of hydroxyapatite on Ti-based alloys, an electrophoresis laboratory installation was used. The preliminary process for chemical activation involved immersion in NaOH (10M) solution for 3 h at a temperature of 60 °C. After activation, the sample was washed in an ultrasound bath with acetone, ethyl alcohol, and water for 1 h. For the deposition of HA thin films, a Consort EV 261 was used to activate the HA particles. The schematic experimental lab cell for the deposition is presented in Figure 1.

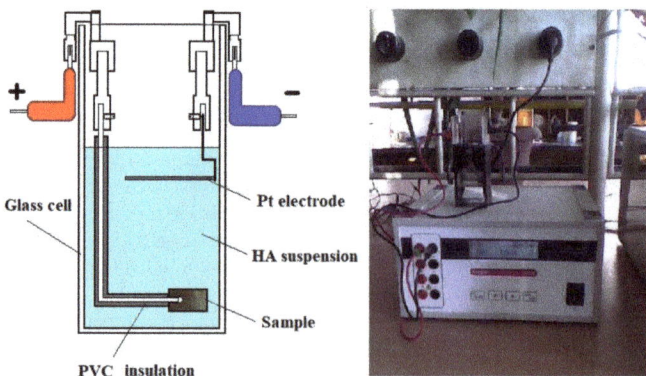

Figure 1. Deposition cell scheme and experimental set-up.

During the process, a voltage of 75 V was applied between the anode (Ti4Al4Zr alloy, S = 3 cm^2) and cathode (platinum, S = 6 cm^2) for 15 min, with a distance of 20 mm between the electrodes. A suspension of 4 g HA powder (0.61 µm average diameter) in 100 mL isopropyl alcohol, stabilized with 1 mL Tween 80 surfactant, was used in the process. After obtaining the growth film, the sample was washed with water, heated in a furnace at 110 °C for 2 h, and calcinated at 800 °C for 2 h.

The surface state of the samples was investigated via electronic microscopy (SEM; VegaTescan LMH II, SE detector, 30 kV electron gun voltage, VegaTC software) with both 2D and 3D images. The resistance to corrosion was determined using the impedance method, also known as EIS.

The EIS spectra are presented as Bode plots of the logarithm of impedance magnitude and the phase angle as a function of frequency. The EIS spectrum was registered in the frequency range of 10-2–10-5 Hz at an alternate potential with 10 MV amplitude using a potentiostat PGZ 301 (VoltaLab 40). Experimental data were converted using EIS file converter software (EISFC150) followed by Z SimWin.

EIS was realized in a solution of simulated blood serum (SBF) but with a concentration multiplied 5-fold (5 × SBF) in order to observe the specific effects in the human body. The measurements were made at potential in an open circuit in a naturally aerated solution.

3. Results

The surface of the D1 and D2 Ti4Al4Zr samples, processed through chemical roughening, were investigated using electron microscopy; Figure 2 presents the surface of the samples after chemical attack: (a,b) sample D1 and (c,d) sample D2. The samples were also analyzed and characterized from a 3D point of view using the analysis software VegaTescan. These results are presented in Table 1. The software analysis was able to detect modifications of the surface different from those obtainable by mechanical means. The microscopy results revealed partial contamination of the surface with impurities following the chemical attack of the surface, although these impurities were not stable on the metallic surface and were easily removed via alcohol washing.

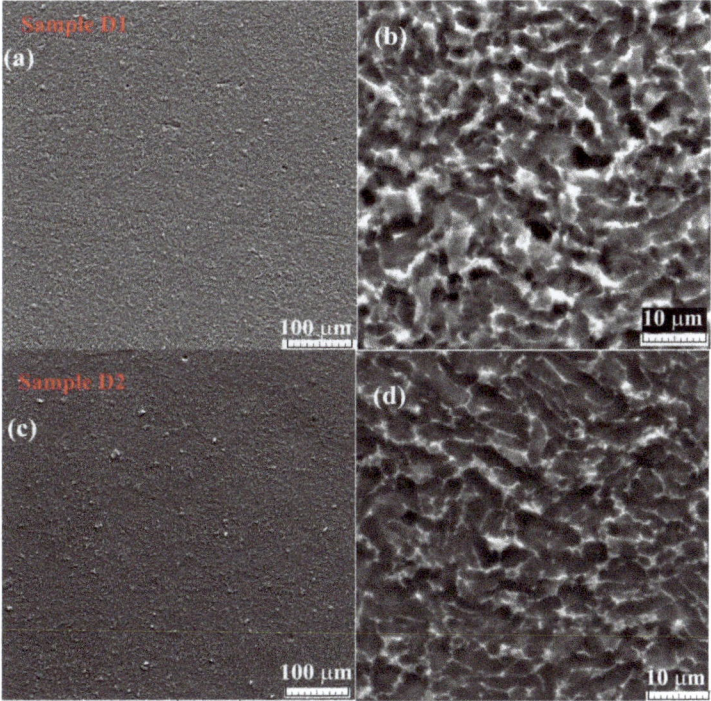

Figure 2. SEM micrographs of the sample surfaces: (**a**) sample D1-500x, (**b**) sample D1-5000x, (**c**) sample D2-500x, and (**d**) sample D2-5000x.

Table 1. Surface characteristics of samples D1 and D2 (values were obtained from Figure 3 using VegaT software).

Sample/Dimensions (from 2D images)	Radius (μm)				Area (μm²)			
	min	med	max	StDev	min	med	max	StDev
D1	1.86	3.19	4.79	0.74	10.83	33.70	72.18	15.87
D2	1.24	2.41	5.44	0.71	4.87	19.91	92.97	13.87

Sample/Dimensions (from 3D images)	H_{min} (μm)	H_{med} (μm)	H_{max} (μm)	StDev (μm)	Minimum variation (ADU)	Maximum variation (ADU)	Differences (ADU)
D1	2.38	3.90	6.33	1.01	60	68.5	8.5
D2	1.08	1.33	1.87	0.21	59.6	63.4	3.8

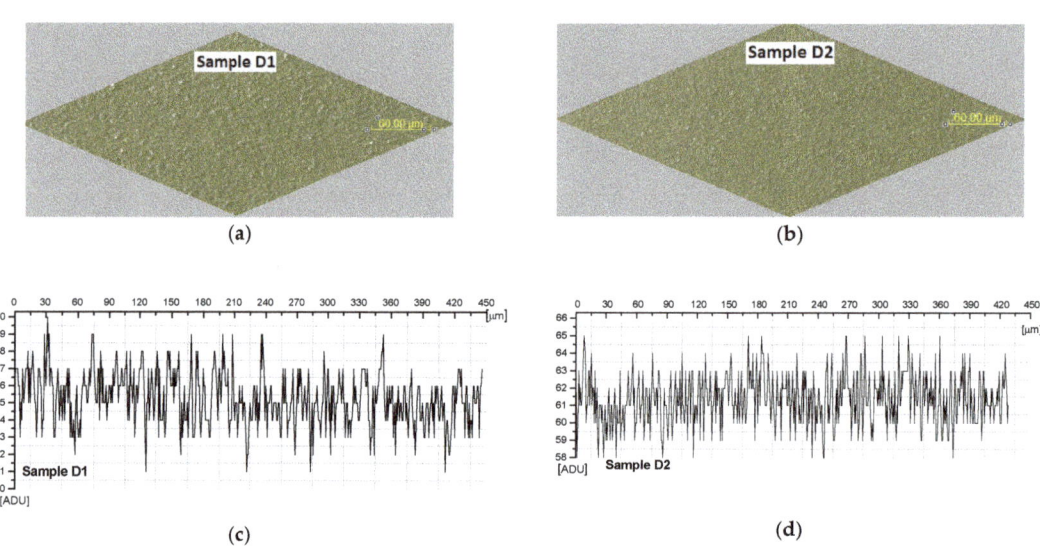

Figure 3. 3D images of sample surfaces: (a) D1 and (b) D2. Surface roughness distribution: (c) D1 and (d) D2.

The chemical analysis of the surfaces did not reveal the contamination of the material surface; the traces of chemical attack were removed via a wash with distilled water. The modifications of the Ti4Al4Zr sample surface through chemical processes can be classified into two types: those that modify the surface at the micron level (D1) and those that modify the surface at the nanometric level (D2).

The layers of HA deposited on the metallic material were of the level of microns (5–10 μm), which led to an influence of its properties at this level. The surfaces engraved at the nanometric level were more appropriate for nanometric layers and those at micron sizes for micron layers, which can cover and homogenize the surface after deposition.

Figure 3 presents the 3D microscopy images of the samples' chemically treated surfaces. Using the 3D microscopes, the performance of the chemical attack methods was assessed. This information is organized in Table 1 (including the minimum, medium, and maximum depths of the traces on the surface).

The effects of the chemical etching were observed on sample D1 at a maximum 6.33 μm depth in an isolated case, with all the other samples having values between 0.81 and 2.4 μm. The standard deviations were at the submicron level and confirmed a good general homogenization of the sample via the chemical processing applied on the surface. Figure 3c,d presents the surface state in terms of the distribution of luminous intensity (surface roughness indicator) on the chemically processed metallic surface. The associated values are shown in Table 1.

Figure 3 also shows a noticeably compact longitudinal distribution of the surfaces modified through a chemical attack in the case of sample D2, while sample D1 showed deeper effects and a greater roughness. The surface state influences the quality of the deposited HA layers through an electrophoretic process based on the correlation between the surface roughness and the layer thickness [21,22].

The SEM microscopy images of the superficial layer of HA deposited using electrophoresis for (a) chemically processed substrate sample D1 at 150× amplification and (b) D2 with details at 1000× amplification are presented in Figure 4. From the images, we observed a good homogeneity of the layer at the macrostructural level and, in some cases (Figure 4b), cracking in certain areas at the microstructural level, especially after the thermal calcination treatment at 800 °C of sample D2, which served to stabilize the layer of deposited hydroxyapatite.

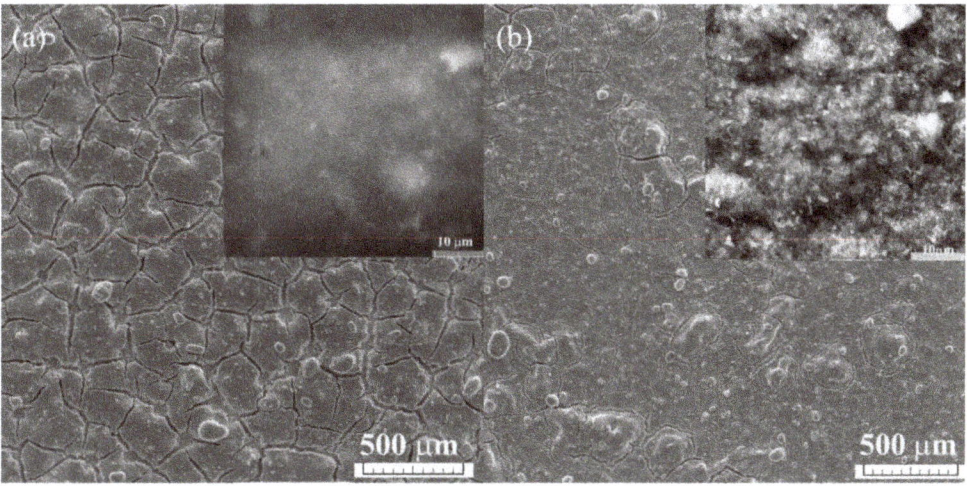

Figure 4. SEM micrographs of superficial HA layer realized through electrophoretic deposition method: (**a**) sample D1 and (**b**) sample D2.

The type of ceramic structure formed during the electrophoresis process of the TiO_2 and HA layers is highlighted in Figure 5, in which the variation of chemical elements on a scratched surface of Ti4Al4Zr are presented. The appearance of a TiO_2 layer formed at the interface between the metal substrate and the HA surface layer was determined. The titanium oxide layer identified by SEM–EDS and confirmed by EIS was a very homogeneous and resistant layer with a compact microstructural appearance and submicron dimension.

Biocompatible micro- to nanostructured covering layers play a crucial role in medical applications for the osseointegration capacity of more resistant metallic implants. Both TiO_2 and HA layers influence the corrosion resistance of the metallic material, with the former improving the corrosion resistance and the latter offering better biocompatibility.

Using the Z SimWin program, it was found that the EIS data obtained for sample D1 could be described by the equivalent circuit presented in Figure 6, together with a Bode diagram and with the value of the parameter's characteristic in the elements of the equivalent circuit [23,24].

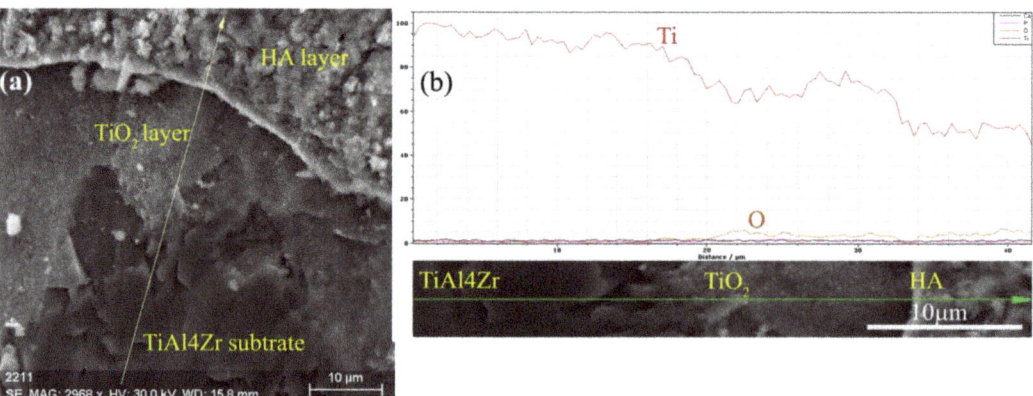

Figure 5. Complex structure identification of HA deposition through line chemical composition mode (**a**) SEM image of analyzed area; (**b**) chemical elements line variations

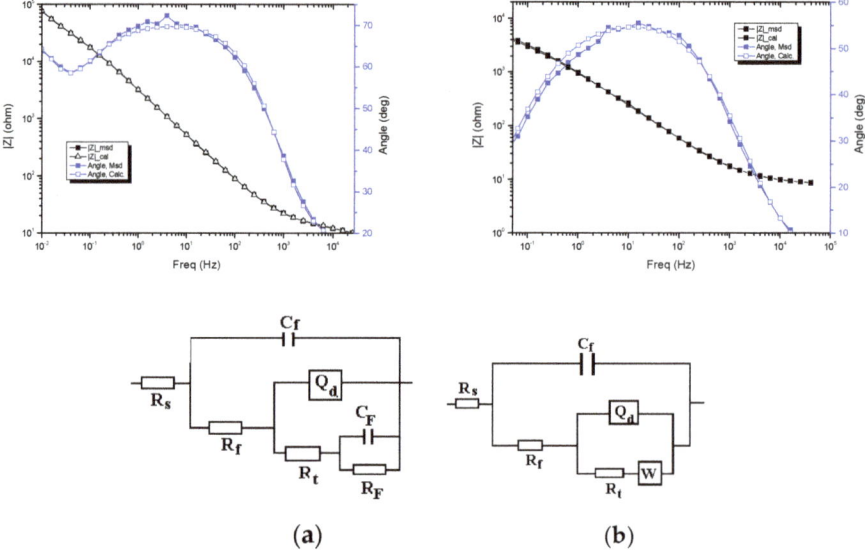

Figure 6. EIS data, Bode diagram, and equivalent circuit obtained for: (**a**) sample D1 and (**b**). sample D2.

In the equivalent circuit, R_s is the resistance of solution, R_f and Cf are the resistance and capacity of the superficial layer, R_t is the resistance to the transfer charge, Q_d is the element of the phase constant associated with the electric double layer (instead of the capacity of the double electrical layer), and the parallel circuit R_f–C_f is the existence of the Faraday process, the origin of which cannot be explained (a Faraday current is the result of an oxidation–reduction reaction that takes place at the surface of the electrode), but it might exist due to the process of a depositing or dissolving reactions that takes place the surface of the electrode. It also could be the result of a depositing or dissolving reaction that was produced during the measurements [25].

The parameter χ^2 from the last column of Table 2 is a measure of accuracy for the model chosen to describe the experimental data to the value $\chi^2 = 9.41 \times 10^{-4}$. The value corresponds to a relative error in measuring the current equal to $\sum = 2\%$, while the process found that this model was satisfactory for interpreting the data. The very small value of superficial flux resistance R_f (almost equal to solution resistance) indicates that this layer had surface defects.

Table 2. The values of the equivalent circuit for D1.

Sample	R_S (Ω cm^2)	C_f (F/cm^2)	R_f (Ω cm^2)	Q_d (S s^n cm^{-2})	n	R_t (Ω cm^2)	C_F (F/cm^2)	R_f (Ω cm^2)	χ^2
D1	10.01	3.51×10^{-6}	8.91	7.1×10^{-5}	0.77	9.8×10^4	8.51×10^{-5}	9.81×10^5	3.96×10^{-4}

Noting that the layer capacity value C_f was at the level of micro-farads (3.51 µf), which denotes a small thickness of the layer (at the level of hundreds of nanometers, this value was high (98.7 Kohm), the current of corrosion was low (in the order of hundreds of nA), and the process of corrosion in this solution was not significant. The value of the exponent n from the expression of the element of constant phase Q_d (n =0.77) was far below 1, corresponding to on ideal condenser, which means that the capacity of the double layer was an imperfect condenser due to the imperfection of the surface alloy support associated with sanding and the nonuniformity of the hydroxyapatite layer deposed through electrophoresis.

The experimental results obtained for sample D2 were processed using the same equivalent circuit as sample D1. The values of the elements of the circuit are shown in Table 3. Analysis of these data indicated that using this circuit to describe the macroscopic structure of the layer deposited on this sample was not acceptable.

Table 3. The values of equivalent circuit for sample D2.

Sample	R_S (Ω cm^2)	C_f (F/cm^2)	R_f (Ω cm^2)	Q_d (S s^n cm^{-2})	n	R_t (Ω cm^2)	C_F (F/cm^2)	R_f (Ω cm^2)	χ^2
D2	8.49	4.22×10^{-6}	4.57	3.01×10^{-5}	0.6	1.0×10^{-7}	3.81×10^{-6}	6.92×10^3	9.41×10^{-4}

It is worth emphasizing the high value of the χ^2 parameter (9.41 × 10^{-4}), representing a relative error of around 3%. If the solution resistance value (R_s) had the same order of dimension as sample D1, we cannot explain why the resistance of the solution would indicate the concentration of the solution. The resistance of the charge transfer had an unacceptable value of around 0. For this reason, the interaction of the metal solution is possible without restriction. The exponent **n** from the expression of Q_d (n =0.6) indicated that the constant phase element could not be assimilated with a capacity of the double layer, being much closer to the 0.5 value characteristic of a corrosion process controlled by diffusion. The Faraday process was also present in this case but of lesser importance.

Due to the imperfection of the model used to process the experimental data obtained for sample D2, another circuit was applied in which the Warburg impedance (w = $\sigma \times ^{1/2}$, where w is the angle frequency, σ is the time constant for diffusion, and x is the Warburg coefficient that expresses the mass transfer through the superficial film, as when the corrosion process is controlled partially by diffusion) was introduced. This equivalent circuit is presented in Figure 6b, together with the Bode diagram and the values of the circuit elements are presented in Table 4.

Table 4. Values of the equivalent circuit for sample D2.

Sample	R_S (Ω cm^2)	C_f (F/cm^2)	R_f (Ω cm^2)	Q_d (S s^n cm^{-2})	n	R_t (Ω cm^2)	W (S s$^{1/2}$ cm^{-2})	χ^2
D2	8.71	4.28×10^{-6}	3.37	2.94×10^{-4}	0.63	6.09×10^3	2.25×10^{-3}	5.54×10^{-4}

In this condition, like those for sample D2, the superficial layers appeared very permeable to the solution (porous layer imbibed with solution) and could not protect the metal. The resistance for the charge transfer (Rt) was much smaller than in the case of sample D1 (16 times smaller), indicating a much more pronounced tendency of interaction between the electrode and solution.

The significance of the other elements in this circuit is the same as that of the previous equivalent circuit. The values of the solution resistance (R_s) and of the superficial film resistance were similar to those obtained with the previous model and were very close to those obtained for the D1 sample.

The Warburg impedance (W) was very low, so the corrosion process was not significantly influenced by the mass transfer but only controlled by the chemical reaction. The layer deposited through electrophoresis on this sample was much more irregular and less protective than that obtained for sample D1.

The cyclic polarization curves were registered in the interval (−100 ... + 1000 ... −1000) mV with a scanning speed of 10 mV/s. The cyclic polarization curves obtained for the two samples are presented in Figure 7a,b.

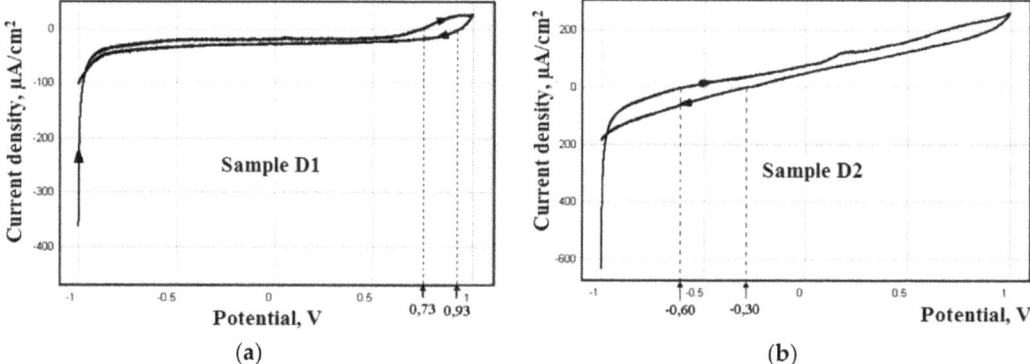

Figure 7. Cyclic polarization curves for samples (**a**) D1 and (**b**) D2.

The cyclic polarization curves obtained differed significantly between samples D1 and D2, and they were in accordance with the results of the electrochemical impedance spectroscopy measurements. Thus, for sample D1, the layer of HA deposited on the alloy surface assured protection against corrosion in the 5 × SBF solution only to overpotentials smaller than 730 mV; above this level, chemical interactions between the metal and solution took place. The corrosion rate was not high; at an overpotential of 1000 mV, the density of the corrosion current was only 27 µA/cm². On the turning curve (the cathode brand of cyclic voltammograms), re-passivation took place at an overpotential of 930 mV.

The layer deposed on sample D2 did not assure protection against corrosion in the 5 × SBF solution. Thus, the corrosion process occurred at negative potentials: I = 0 at potential E = −600 mV (E0 = −600 mV represents the corrosion potential for this sample). At higher potentials (>E0), the corrosion speed grew proportionally with the overpotential, so that at the potential 1000 mV, the corrosion current reached the value of 264 µA/cm², which was 10 times higher than that of sample D1. On the turning curve, re-passivation took place at potentials under −300 mV. In these conditions, we conclude that the layer deposited on sample D1 was protective only at overpotentials under 730 mV, while the layer deposited on sample D2 assured no protection to corrosion.

4. Conclusions

The superficial layer obtained on the Ti4Al4Zr alloy after deposition and calcination was represented by a compact layer of TiO_2 that was very well adhered, and a less compact, nonhomogeneous, semiporous layer made of hydroxyapatite deposited through electrophoresis. This combination ensures reliable protection for the alloy's surface as well as good biocompatibility. The layer deposited on sample D2 did not ensure protection against corrosion in a 5 × SBF solution. In these conditions, it can be concluded that the bilayer (TiO$_2$–HA) on sample D1 was protective only at overpotentials below 730 mV, while the layer deposited on sample D2 did not ensure representative protection against corrosion. This difference in the layer's effect on corrosion resistance is attributed to the state of the metallic material surface.

Author Contributions: Conceptualization, R.C. and P.V.; methodology, N.C.; software, R.C. and M.P.; validation, R.C., P.V., and M.P.; formal analysis, I.Ș.; investigation, N.C. and G.Z.; resources, N.I.; data curation, A.S.; writing—R.C., N.C., and M.P.; writing—review and editing, P.V. and M.P.; visualization, G.Z.; supervision, N.C.; project administration, I.Ș.; funding acquisition, N.C. Please turn to the CRediT taxonomy for the term explanation. Authorship must be limited to those who have contributed substantially to the work reported. All authors have read and agreed to the published version of the manuscript.

Funding: This research was funded by the Romanian Ministry of Education and Research, CNCS-UEFISCDI, project number PN-III-P1-1.1-TE-2019-1921, within PNCDI III.

Institutional Review Board Statement: Not applicable.

Informed Consent Statement: Not applicable.

Conflicts of Interest: The authors declare no conflict of interest.

References

1. Kaur, M.; Singh, K. Review on titanium and titanium-based alloys as biomaterials for orthopedic applications. *Mater. Sci. Eng. C* **2019**, *102*, 844–862. [CrossRef]
2. Luo, H.; Wu, Y.; Diao, X.; Shi, W.; Feng, F.; Qian, F.; Umeda, J.; Kondoh, K.; Xin, H.; Shen, J. Mechanical properties and biocompatibility of titanium with a high oxygen concentration for dental implants. *Mater. Sci. Eng. C* **2020**, *117*, 111306. [CrossRef]
3. Ionita, D.; Grecu, M.; Ungureanu, C.; Demetrescu, I. Modifying the TiAlZr biomaterial surface with coating, for a better anticorrosive and antibacterial performance. *Appl. Surf. Sci.* **2011**, *257*, 9164–9168. [CrossRef]
4. Ionita, D.; Grecu, M.; Ungureanu, C.; Demetrescu, I. Antimicrobial activity of the surface coatings on TiAlZr implant biomaterial. *J. Biosci. Bioeng.* **2011**, *112*, 630–634. [CrossRef]
5. Ungureanu, C.; Pirvu, C.; Mindroiu, M.; Demetrescu, I. Antibacterial polymeric coating based on polypyrrole and polyethylene glycol on a new alloy TiAlZr. *Prog. Org. Coatings* **2012**, *75*, 349–355. [CrossRef]
6. Narayanan, R.; Seshadri, S.K. Point defect model and corrosion of anodic oxide coatings on Ti–6Al–4V. *Corros. Sci.* **2008**, *50*, 1521–1529. [CrossRef]
7. Demetrescu, I. Passive and Bioactive Films on Implant Materials and their Efficiency in Regenerative Medicine. *Mol. Cryst. Liq. Cryst.* **2008**, *486*, 110–119. [CrossRef]
8. Wu, C.; Tang, Y.; Mao, B.; Zhao, K.; Cao, S.; Wu, Z. Rapid apatite induction of polarized hydrophilic HA/PVDF bio-piezoelectric coating on titanium surface. *Surf. Coatings Technol.* **2021**, *405*, 126510. [CrossRef]
9. Izquierdo, J.; Bolat, G.; Cimpoesu, N.; Trinca, L.C.; Mareci, D.; Souto, R.M. Electrochemical characterization of pulsed layer deposited hydroxyapatite-zirconia layers on Ti-21Nb-15Ta-6Zr alloy for biomedical application. *Appl. Surf. Sci.* **2016**, *385*, 368–378. [CrossRef]
10. Mareci, D.; Cimpoeşu, N.; Popa, M.I. Electrochemical and SEM characterization of NiTi alloy coated with chitosan by PLD technique. *Mater. Corros.* **2012**, *63*, 985–991. [CrossRef]
11. Istrate, B.; Rau, J.V.; Munteanu, C.; Antoniac, I.V.; Saceleanu, V. Properties and in vitro assessment of ZrO2-based coatings obtained by atmospheric plasma jet spraying on biodegradable Mg-Ca and Mg-Ca-Zr alloys. *Ceram. Int.* **2020**, *46*, 15897–15906. [CrossRef]
12. Vargas-Becerril, N.; Sánchez-Téllez, D.; Zarazúa-Villalobos, L.; González-García, D.; Álvarez-Pérez, M.; de León-Escobedo, C.; Téllez-Jurado, L. Structure of biomimetic apatite grown on hydroxyapatite (HA). *Ceram. Int.* **2020**, *46*, 28806–28813. [CrossRef]
13. Zu, X.T.; Liu, Y.Z.; Lian, J.; Liu, H.; Wang, Y.; Wang, Y.H.; Wang, L.M.; Ewing, R.C. Surface modification of a Ti–Al–Zr alloy by niobium ion implantation. *Surf. Coat. Tech.* **2006**, *201*, 3756–3760. [CrossRef]
14. Liu, Y.; Zu, X.; Qiu, S.; Wang, L.; Ma, W.; Wei, C. Surface characterization and corrosion resistance of Ti–Al–Zr alloy by niobium ion implantation. *Nucl. Instrum. Methods Phys. Res. Sect. B* **2006**, *244*, 397–402. [CrossRef]
15. Bansal, P.; Singh, G.; Sidhu, H.S. Improvement of surface properties and corrosion resistance of Ti13Nb13Zr titanium alloy by plasma-sprayed HA/ZnO coatings for biomedical applications. *Mater. Chem. Phys.* **2021**, *257*, 123738. [CrossRef]
16. Gradinariu, I.; Stirbu, I.; Gheorghe, C.A.; Cimpoesu, N.; Agop, M.; Cimpoeşu, R.; Popa, C. Chemical properties of hydroxyapatite deposited through electrophoretic process on different sandblasted samples. *Mater. Sci.* **2014**, *32*, 578–582. [CrossRef]

17. Kamitakahara, M.; Kimura, K.; Ioku, K. Synthesis of nanosized porous hydroxyapatite granules in hydrogel by electrophoresis. *Colloids Surf. B Biointerfaces* **2012**, *97*, 236–239. [CrossRef]
18. John, A.S.; Sidek, M.M.; Thang, L.Y.; Sami, S.; Tey, H.Y.; See, H.H. Online sample preconcentration techniques in nonaqueous capillary and microchip electrophoresis. *J. Chromatogr. A* **2021**, *1638*, 461868. [CrossRef] [PubMed]
19. Hosseini, M.R.; Ahangari, M.; Johar, M.H.; Allahkaram, S.R. Optimization of nano HA-SiC coating on AISI 316L medical grade stainless steel via electrophoretic deposition. *Mater. Lett.* **2021**, *285*, 129097. [CrossRef]
20. Zirom-Titanium. Available online: https://www.zirom-titanium.com/ (accessed on 10 October 2020).
21. Stirbu, I.; Vizureanu, P.; Cimpoesu, R.; Dascălu, G.; Gurlui, S.O.; Bernevig, M.; Benchea, M.; Cimpoeşu, N.; Postolache, P. Advanced metallic materials response at laser excitation for medical applications. *J. Optoelectron. Adv. M.* **2015**, *17*, 1179–1185.
22. Afiq Harun, M.; Zamree Abd Rahim, S.; Nasir Mat Saad, M.; Fathullah Ghazali, M. Warpage analysis on front panel housing using response surface methodology (RSM). *Eur. J Mater. Sci. Eng.* **2016**, *1*, 9–18.
23. Cimpoeşu, N.; Săndulache, F.; Istrate, B.; Cimpoeşu, R.; Zegan, G. Electrochemical Behavior of Biodegradable FeMnSi–MgCa Alloy. *Metals* **2018**, *8*, 541. [CrossRef]
24. Istrate, B.; Munteanu, C.; Lupescu, S.; Chelariu, R.; Vlad, M.D.; Vizureanu, P. Electrochemical Analysis and In Vitro Assay of Mg-0.5Ca-xY Biodegradable Alloys. *Materials* **2020**, *13*, 3082. [CrossRef] [PubMed]
25. Cimpoeşu, R.H.; Pompilian, G.O.; Baciu, C.; Cimpoeşu, N.; Nejneru, C.; Agop, M.; Gurlui, S.; Focşa, C. Pulsed laser deposition of poly (L-Lactide) acid on nitinol substrate. *Optoelectron. Adv. Mat.* **2010**, *4*, 2148–2153.

Article

Magnetic Abrasive Finishing of Beta-Titanium Wire Using Multiple Transfer Movement Method

Sung Sik Nam [1], Jeong Su Kim [2] and Sang Don Mun [1,2,*]

[1] Department of Mechanical Design Engineering, Jeonbuk National University, 567, Baekje-daero, Deokjin-gu, Jeonju-si, Jeollabuk-do 54896, Korea; 201950147@jbnu.ac.kr

[2] Department of Energy Storage/Conversion Engineering of Graduate School, Jeonbuk National University, Jeonju, Jeollabuk-do 54896, Korea; kjs1592@jbnu.ac.kr

* Correspondence: msd@jbnu.ac.kr

Received: 25 August 2020; Accepted: 22 September 2020; Published: 25 September 2020

Abstract: Titanium is often used in various important applications in transportation and the healthcare industry. The goal of this study was to determine the optimum processing of magnetic abrasives in beta-titanium wire, which is often used in frames for eyeglasses because of its excellent elasticity among titanium alloys. To check the performance of the magnetic abrasive finishing process, the surface roughness (Ra) was measured when the specimen was machined at various rotational speeds (700, 1500, and 2000 rpm) in the presence of diamond paste of various particle sizes (0.5, 1, and 3 μm). We concluded that the surface roughness (Ra) was the best at 2000 rpm, 1 μm particle size, and 300 s processing time, and the surface roughness of β-titanium improved from 0.32 to 0.05 μm. In addition, the optimal conditions were used to test the influence of the finishing gap, and it was found that the processing power was superior at a gap of 3 mm than at 5 mm when processing was conducted for 300 s.

Keywords: magnetic abrasive finishing; FEMM; surface roughness; multi-feed movement; beta-titanium wire; EDS elemental mapping; atomic force microscope

1. Introduction

Materials with good mechanical properties are needed in high-tech applications. Therefore, many studies are being conducted to achieve high accuracy and improve the quality of products, particularly with regard to surface properties. Many machining methods have been adopted for high-precision processing technology [1,2]. However, the surface accuracy results obtained are not satisfactory in many industries. Wire titanium comes in the form of pure titanium alloy, which has been widely used in eyeglasses frames because of its good elasticity, resilience, and ability to bend to fit the shape of the face. Beta titanium is lighter than regular titanium and can be a good alternative material for weight reduction because it is stronger, resistant to heat, and shows less deformation and deterioration caused by salt [3]. Thus, it has been applied in clinical situations, for example, in braces for teeth. However, beta-titanium alloy is difficult to machine due to its low thermal conductivity. The magnetic abrasive finishing (MAF) process is an ultra-precision machining technique which uses magnetic fields to process hard-to-work pieces [4–6]. Heng et al. [7] applied the magnetic abrasive finishing process to achieve a high surface quality of magnesium alloy bar (Ø3 mm × 50 mm). His result showed that the value of surface roughness was reduced 0.02 μm after the finishing process within 20 s. Chang et al. [8] used the magnetic abrasive finishing process for cylindrical surface finishing of SKD11 steel material (Ø15 mm × 80 mm). His result showed that after the finishing process, the best surface roughness (Ra) of 0.042 μm was achieved by 180 μm steel grit abrasive.

However, workpieces made of small diameter beta-titanium alloy wire may experience fractures and micro-cracks due to the high pressure of the magnetic abrasive finishing process [9–12].

Therefore, the magnetic abrasive finishing process must be employed with new processing parameters such as a wire moving system (WMS), a wire vibration system (WVS), or a combination of WMS and WVS.

We conducted a study to identify the optimal conditions for beta-titanium workpieces using a magnetic abrasive finishing process with multiple transfer movement methods. The experiments were carried out with various abrasive particle sizes, rotational speeds, finishing gaps, and multiple transport motion methods under the chosen processing conditions.

2. Experimental Equipment and Methods

2.1. Experimental Equipment

Figures 1 and 2 show the experimental setups of the magnetic abrasive finishing process and the finishing equipment for beta-titanium wire alloys, respectively. The magnetic abrasive finishing (MAF) equipment uses a rotational magnetic field and consists of a stepping motor and a system controller. These allow forward rotation and reverse rotation with ultra-high precision to improve the surface accuracy of machined beta-titanium alloy. This equipment was designed to allow the wire workpieces to reciprocate inside a flexible magnetic abrasive brush. MAF equipment for beta-titanium wire alloy processing is divided into three main parts: (i) the wire moving system (WMS), (ii) the rotational magnetic field system (RMS), and (iii) the wire vibration system (WVS). First, the wire moving system (WMS) consists of two sensors and a drive spool, stepping motor, belt, and power supply. The WMS reciprocates wires left and right inside the rotational magnetic field. Secondly, the RMS system consists of a belt, rpm controller, pulley, (Nd-Fe-B) permanent magnet, steel yoke, steel chuck, belt, program controller, stepping motor, and the unbonded magnetic abrasive materials. Permanent magnets (Nd-Fe-B) are used because they are small in volume and can generate high magnetic flux density, which can improve the resulting surface roughness [13–16]. Finally, the wire vibration system (WVS) consists of an electronic slider, a power supply, and a programmable controller. This method generates vibration in a rotational magnetic field of 8 Hz with 2 mm amplitude. The vibration of the magnetic field rotating system is used for increasing the number of finishing times, allowing the magnetic smoothing to proceed efficiently on the workpiece surface. The specimen used in this study is β-titanium wire with a diameter of 0.5 mm and a length of 50 mm. Figure 3 shows a schematic view of the device used to generate a rotational magnetic field for processing beta-titanium wire. Figure 3a shows the 3D view of rotational magnetic field device. Figure 3b shows the 2D dimensional view of the rotational magnetic field device used in this experiment. The finishing device consists of two sets of Nd-Fe-B magnets (20 mm × 10 mm × 5 mm), Al 6063 chuck, AISI 1018 steel yoke, and two sharp edges of AISI 1018 steel magnetic poles.

Appl. Sci. **2020**, *10*, 6729

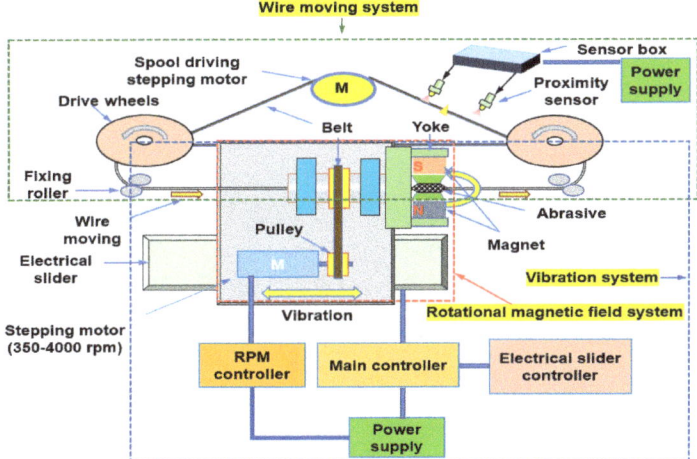

Figure 1. Experimental setup of magnetic abrasive finishing process for beta-titanium wire.

Figure 2. Magnetic abrasive finishing equipment for beta-titanium wire using Nd-Fe-B permanent magnet.

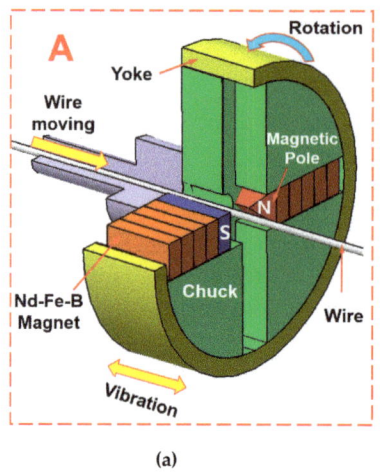 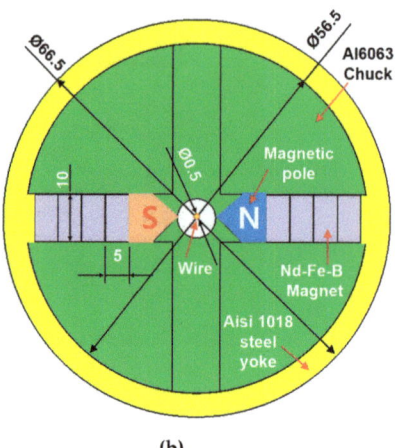

Figure 3. Schematic view of rotating magnetic field generator for beta-titanium wire: (**a**) 3D view of rotational magnetic field device and (**b**) 2D dimensional view of the rotational magnetic field device.

2.2. Beta-Titanium Wire Material

Compared with stainless steel wires, beta-titanium wire material has lower force magnitudes, lower elastic modulus, higher spring back (maximum elastic deflection), lower yield strength, and good ductility, weldability, and corrosion resistance. Its formability and weldability are advantages over Ni-Ti alloys [17]. Due to its excellent mechanical properties, it has been used in medical applications, sport applications, and as frames for eyeglasses. Therefore, beta-titanium wire material was used in this study as the workpiece with a diameter of 0.5 mm and a length of 50 mm. The chemical composition and mechanical properties of the specimen are shown in Tables 1 and 2.

Table 1. Chemical composition of β-titanium wire.

Component	N	C	H
Chemical composition (Max%)	0.03	0.08	0.015
Component	O	Fe	Ti
Chemical composition (Max%)	0.18	0.2	Remainder

Table 2. Mechanical properties of β-titanium wire.

Mechanical Properties	Matrix
Tensile strength (MPa)	345
Yield strength (MPa)	220
Elongation (%)	35
Hardness (HV)	115
Elastic modulus (GPa)	115
Density (g/m^3)	4.51

2.3. Experimental Methods

The experimental conditions are as shown in Table 3. The beta-titanium wire workpiece was 0.5 mm in diameter, 50 mm in length, and 0.33 µm Ra in surface roughness. Unbonded magnetic abrasive material was used in this study, which is a mixture of electrolytic iron powder (1.2 g, Fe #200), diamond paste (0.6 g), carbon powder (0.02 g), and light oil (0.3 mL). To find the optimal conditions for processing, we studied three sample β-titanium wire rotation speeds (700, 1500, 2000 rpm) and three diamond particle sizes (0.5, 1, and 3 μm). The process was conducted for 300 s of total finishing time at each condition. Ultrasonic cleaning and drying were performed every 60 s of finishing time to allow measurement of the correct surface roughness (Ra) of the β-titanium wire workpiece. In addition, the optimal conditions obtained from the previous two experiments (i.e., rotation speed, abrasive particle size) were applied with different finishing gaps (3 and 5 mm). The distribution of magnetic flux density was analyzed using a finite element method magnet (FEMM). To understand the change of surface roughness after the finishing process, the average surface roughness (Ra) values were measured every 60 s; three times at different positions by a surface roughness tester (Mitutoyo SJ-400, Japan).

Table 3. Experimental conditions.

Workpiece material	β-titanium wire (L = 50 mm, D = 0.5 mm)
Electrolytic iron powder	1.2 g (#200)
Diamond paste (PCD)	(0.5, 1, and 3 μm) 0.6 g
CNT particle	0.02 g
Lubricant	0.3 mL (light oil)
Magnet	Nd-Fe-B permanent magnet (Size: 20 mm × 10 mm × 10 mm)
Magnetic pole vibration	Amplitude: 2 mm
Finishing gap	3 mm, 5 mm
Rotational speed	700, 1500, 2000 rpm
Finishing time	0, 60, 120, 180, 240, 300 s
Vibration frequency	8 Hz
Feed rate	80 mm/min

3. Experimental Results and Analysis

3.1. Influence of Particle Size

Figure 4 shows the changes in surface roughness (Ra) as a function of finishing time for various magnetic abrasive particle sizes. To determine the effect of changes in magnetic abrasive particle sizes (0.5, 1, 3 μm) we performed experiments with an 8 Hz vibration rotational magnetic field at a rotational speed of 1500 rpm. The smallest (0.5 μm) grain size abrasive material was found to have the best effect followed by 1 and 3 μm magnetic abrasive particles. After the finishing process, the surface roughness of the workpieces were 0.12, 0.12, and 0.15 μm for particle sizes of 0.5, 1, and 3 μm, respectively. Thus, smaller magnetic abrasive grain sizes produced better results in terms of the surface roughness.

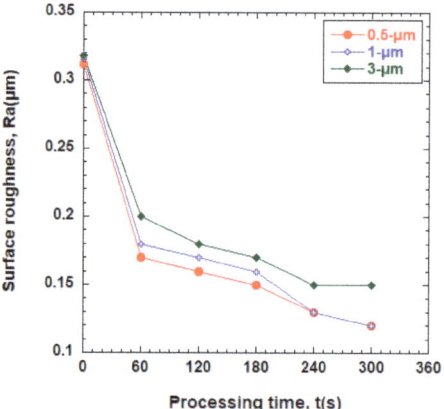

Figure 4. Surface roughness (Ra) vs. processing time for different diamond paste sizes (vibration: 8 Hz; rotation: 1500 rpm; sharp edge; gap: 3 mm).

3.2. Influence of Rotational Speed

To determine the optimum finishing performance as a function of the rotational speed of the workpiece, three rotational speeds were investigated: 700, 1500, and 2000 rpm. For these experiments, the particle size was 1 μm and the magnetic pole shape was a sharp edge. Figure 5 shows the changes in surface roughness (Ra) according to different rotational speeds. The results showed that all the surface roughness values continually improved from 0 s to 300 s of finishing time. The surface roughness (Ra) was improved to 0.18, 0.15, and 0.09 μm at 700, 1500, and 2000 rpm, respectively. Thus, we confirmed that increasing the rotational speed improved the surface roughness (Ra) of the wire workpiece. This can be explained by the fact that increasing the rotational speed of the magnetic field can increase the relative motion between the workpiece surface and magnetic abrasive particles.

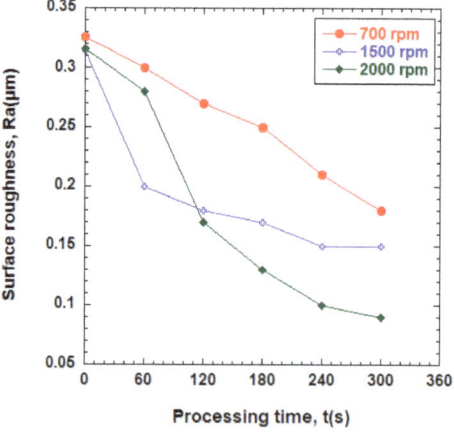

Figure 5. Surface roughness (Ra) vs. processing time as a function of rotational speed (vibration: 8 Hz; abrasive: 1 μm; sharp edge; gap: 3 mm).

3.3. Influence of Clearance Distance

To determine the optimum clearance distance of the workpiece, the magnetic field distribution at the finishing zone was measured using an FEMM. In addition, the chemical composition of beta-titanium wire was analyzed by EDS and the density was verified by mapping. Before the finishing process, the clearance distances between the two magnetic poles were set to 3 and 5 mm. Figure 6 shows the result of the FEMM analysis; the results obtained at the finishing zone were 2.82347 magnetic flux density (T) for 3 mm clearance and 2.56956 T for 5 mm clearance. Figure 7 shows the relationship between magnetic flux density (T) and distance from the magnetic pole. Figure 8 shows the changes in surface roughness (Ra) and processing time for different finishing gaps. The finishing conditions were conducted under the optimal conditions (rotation: 2000 rpm; abrasive: 1 µm; sharp-edged magnetic pole, total finishing time: 300 s). The results showed that surface roughness continually improved from 0 s to 300 s of finishing time. The surface roughness (Ra) improved to 0.08 µm for a gap of 3 mm and 0.21 µm for a gap of 5 mm. This showed that the 3 mm finishing gap was much better than a 5 mm gap. Thus, a smaller finishing gap distance can provide better surface roughness (Ra) on the beta-titanium wire workpiece. According to Figure 6, a higher magnetic flux density was obtained when using smaller finishing gaps. During the finishing process, the higher magnetic flux density generated higher magnetic force, which acted strongly on the magnetic abrasive particles, resulting in a workpiece with better surface quality.

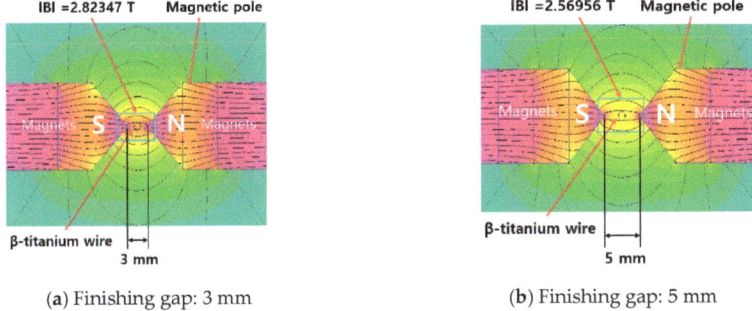

(a) Finishing gap: 3 mm (b) Finishing gap: 5 mm

Figure 6. Front view of sharp-edged magnetic field analyzed by FEMM.

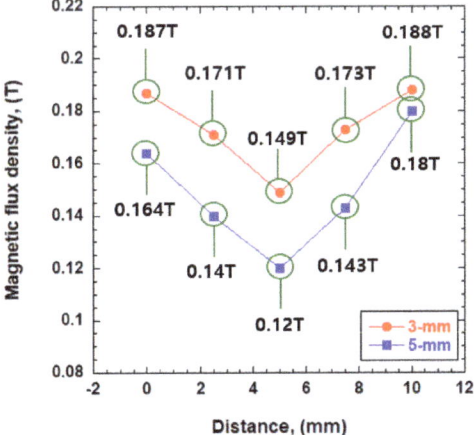

Figure 7. Magnetic flux density (T) vs. distance for different finishing gaps.

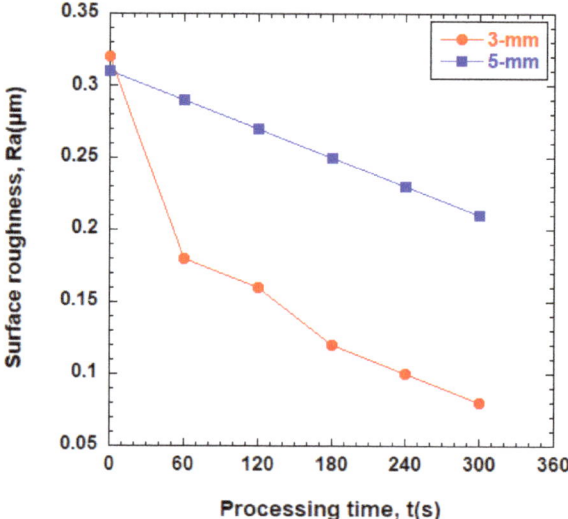

Figure 8. Surface roughness (Ra) vs. processing time for different finishing gaps (vibration: 8 Hz; abrasive: 1 µm; rotation: 2000 rpm; sharp edge magnetic pole).

The surface conditions of beta-titanium wire before and after finishing were investigated using an optical microscope (IMS-M-345) (see Figure 9). Figure 9a shows the initial surface condition before the finishing process. As shown in Figure 9a, the multiple grooves and original scratches can be found everywhere on the initial surface and surface roughness Ra is 0.32 µm. Figure 9b shows the surface condition after the finishing process. From Figure 9b, the surface condition of beta-titanium wire is smoother than the initial surface condition and the multiple grooves were mostly removed. A surface roughness (Ra) after the finishing process is 0.05 µm.

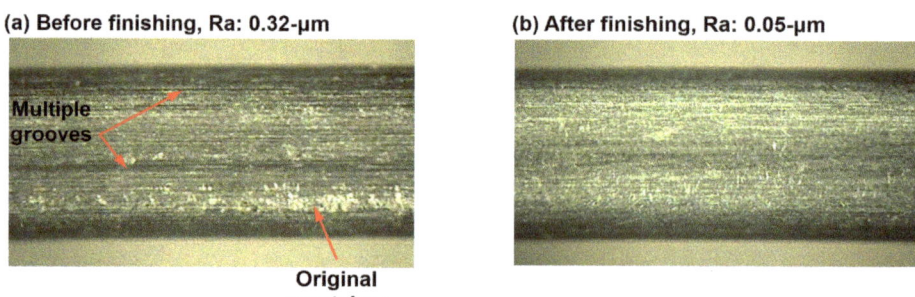

Figure 9. Micro-images of surface roughness before and after finishing: (**a**) before finishing (Ra = 0.32) and (**b**) after finishing (Ra = 0.05).

Figure 10 shows EDS chemical component analysis graphs for beta-titanium wire before and after processing. Figure 10a shows the chemical components of the material detected on the surface of the beta-titanium wire alloy workpiece prior to the finishing process: 55.34% Ti, 26.68% O, 17.85% C, and 0.12% Fe. After the finishing process, Figure 10b shows a composition of 16.40% C, 21.68% O, 61.69% Ti, and 0.23% Fe. Ti and Fe increased after processing, and O and C were lower. However, no other components were found on the surface of the workpiece after the finishing process.

This confirmed that the component magnetic abrasive tools did not affect the surface of the workpiece after the finishing process. Figure 11 shows the EDS chemical composition maps on the surface of the workpiece before and after the finishing process with optimal conditions. Table 4 shows the chemical composition of β-titanium wire before and after finishing.

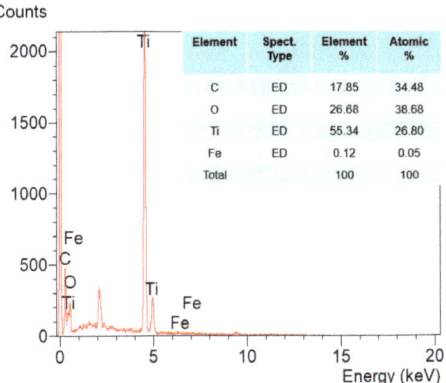

(a) Before the magnetic abrasive finishing process (Ra = 0.32 μm).

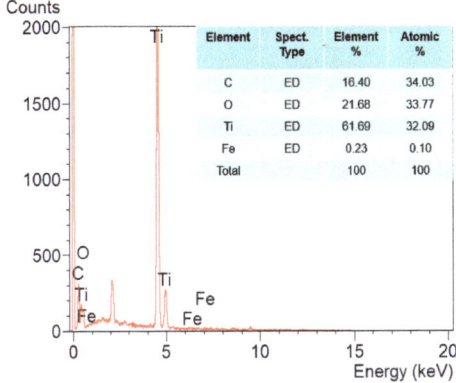

(b) After the magnetic abrasive finishing process (Ra = 0.08 μm).

Figure 10. EDS graphs of the workpiece before and after the magnetic abrasive finishing process (under optimal conditions: 1 μm, 3 mm, 2000 rpm, 300 s).

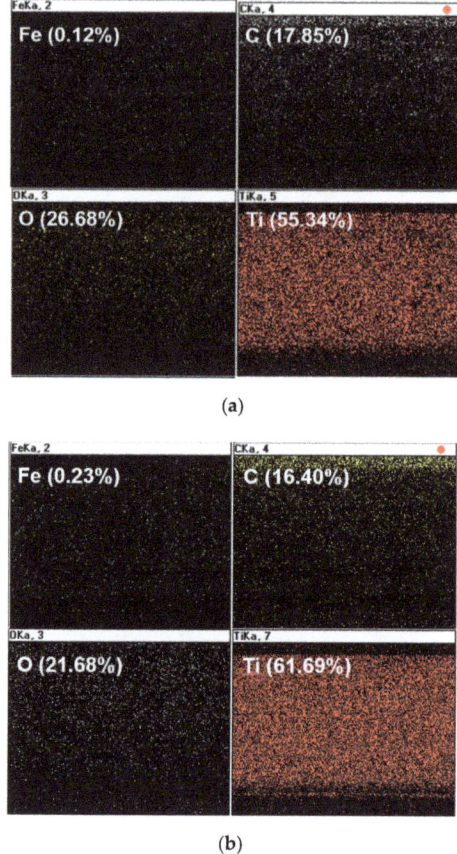

Figure 11. EDS chemical composition mapping of before and after processing the workpiece: (**a**) before finishing Ra = 0.32 and (**b**) after finishing Ra = 0.08.

Table 4. Chemical composition of β-titanium wire before and after finishing.

Element	Chemical Composition (%) before Processing	Chemical Composition (%) after Processing
C	17.85	16.40
O	26.68	21.68
Ti	55.34	61.69
Fe	0.12	0.23
Total	100	100

3.4. Influence of Multiple Transport Movement

Experiments were conducted in three different ways to determine the processing characteristics associated with multiple transport movement. The optimal conditions (2000 rpm; 1 μm; sharp edge, finishing gap: 3 mm) obtained from the previous experiments were used, and the surface characteristics of the wire before and after processing were analyzed using the atomic force microscopy (AFM) 3D-micro images. Figure 12 shows the changes in surface roughness as a function of processing time for different transfer motion methods. In this experiment, the vibration factor was modified in the multi-transportation system, but when finishing time increased to 300 s, the processing performance was improved, and the surface roughness of the workpiece improved to 0.08 µm. The second experiment was conducted using only the feed rate provided by the rotational spools. In this case, the value of Ra was improved to 0.1 µm, which was not as good as the vibration single movement. Finally, vibration and feed rate were both employed at the same time. These results showed the best performance in terms of the surface roughness, showing improvement from 0.32 µm Ra to 0.05 µm Ra. Figure 13 shows the AFM 3D images of the workpiece before and after the various finishing processes: (a) before finishing, (b) vibration with feed rate (8 Hz, 80 mm/min), (c) vibration (8 Hz), (d) feed rate (80 mm/min). As shown in Figure 13, the initial surface of the workpiece was improved in all the conditions. However, better surface conditions were obtained using vibration with feed rate (see Figure 13b).

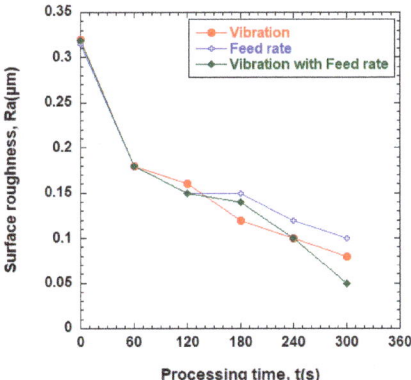

Figure 12. Surface roughness (Ra) vs. finishing time for multi-feed movement (vibration: 8 Hz; abrasive: 1 µm; rotation: 2000 rpm; finishing time: 300 s; sharp edge).

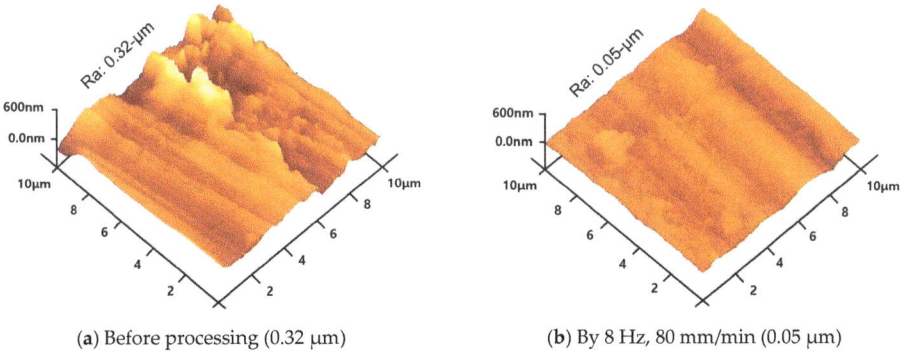

(**a**) Before processing (0.32 µm) (**b**) By 8 Hz, 80 mm/min (0.05 µm)

Figure 13. *Cont.*

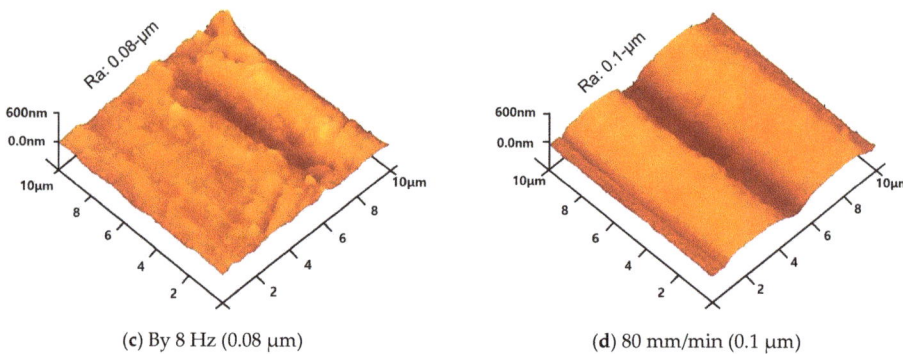

(c) By 8 Hz (0.08 μm) (d) 80 mm/min (0.1 μm)

Figure 13. Atomic force microscopy (AFM) 3D image of the workpiece before and after finishing process according to multi-transportation system employed.

4. Conclusions

Experiments were conducted to find the optimal processing conditions for beta-titanium. The conclusions are as follows:

1. When the grain size of magnetic abrasive material changed at a fixed speed of 1500 rpm, the surface roughness (Ra) was improved to 0.05 μm.
2. When the rotational speed was changed and the particle size was fixed at 1 μm, the best results were obtained at a speed of 2000 rpm.
3. Better finishing was observed at a finishing gap of 3 mm than 5 mm.
4. In all conditions, AFM surface roughness measurements showed that the processed material had a smoother surface than before the machining.
5. When finishing using a multi-transfer motion method under optimal conditions, the processing effects were best to worst in the order: vibration with feed rate, vibration only and feed rate only.
6. Finally, we found that the magnetic abrasive finishing process using a Nd-Fe-B rare earth permanent magnet showed the best effect when the rotational speed was 2000 rpm and 1 μm abrasive material was used with a 3 mm finishing gap and a multi-transfer motion method.

Author Contributions: Design experiment and conceptualization, S.S.N.; methodology and performed experiments, S.S.N. and J.S.K.; investigation and editing, S.D.M.; writing paper, S.S.N. All authors have read and agreed to the published version of the manuscript.

Funding: This research was funded by NATIONAL RESEARCH FOUNDATION (NRF) of Korea in 2016, 2019 (Research Project No. 2016R1D1A1B03932103, 2019R1F1A1061819).

Acknowledgments: The authors appreciate the support from National Research Foundation of Korea (Project No. 2016R1D1A1B03932103, 2019R1F1A1061819) and appreciate the support from Ultra-Precision Finishing Lab for supporting the valuable discussion and guidance on the surface finishing technique.

Conflicts of Interest: The authors have no conflict of interest to declare.

References

1. Choi, S.-Y.; Hwang, C.-U.; Kwon, D.-G. Analysis of Machined Surface Morphology According to Changes of Surface Condition in Micro Particle Blasting. *Korean Soc. Manuf. Process. Eng.* **2018**, *17*, 70–75. [CrossRef]
2. Zou, Y.; Xie, H.; Dong, C.; Wu, J. Study on complex micro surface finishing of alumina ceramic by the magnetic abrasive finishing process using alternating magnetic field. *Int. J. Adv. Manuf. Technol.* **2018**, *97*, 2193–2202. [CrossRef]
3. Kaur, M.; Singh, K. Review on titanium and titanium based alloys as biomaterials for orthopaedic applications. *Mater. Sci. Eng. C* **2019**, *102*, 844–862. [CrossRef] [PubMed]

4. Heng, L.; Yin, C.; Han, S.H.; Song, J.H.; Mun, S.D. Development of a New Ultra-High-Precision Magnetic Abrasive Finishing for Wire Material Using a Rotating Magnetic Field. *Materials* **2019**, *12*, 312. [CrossRef] [PubMed]
5. Wu, J.; Zou, Y.; Sugiyama, H. Study on finishing characteristics of magnetic abrasive finishing process using low-frequency alternating magnetic field. *Int. J. Adv. Manuf. Technol.* **2015**, *85*, 585–594. [CrossRef]
6. Singh, G.; Singh, A.K.; Garg, P. Development of magnetorheological finishing process for external cylindrical surfaces. *Mater. Manuf. Process.* **2017**, *32*, 581–588. [CrossRef]
7. Heng, L.; Yang, G.E.; Wang, R.; Kim, M.S.; Mun, S.D. Effect of carbon nano tube (CNT) particles in magnetic abrasive finishing of Mg alloy bars. *J. Mech. Sci. Technol.* **2015**, *29*, 5325–5333. [CrossRef]
8. Chang, G.-W.; Yan, B.-H.; Hsu, R.-T. Study on cylindrical magnetic abrasive finishing using unbonded magnetic abrasives. *Int. J. Mach. Tools Manuf.* **2002**, *42*, 575–583. [CrossRef]
9. Parameswari, G.; Jain, V.K.; Ramkumar, J.; Nagdeve, L. Experimental investigations into nanofinishing of Ti6Al4V flat disc using magnetorheological finishing process. *Int. J. Adv. Manuf. Technol.* **2017**, *100*, 1055–1065. [CrossRef]
10. Choi, S.-Y.; Kwon, D.K. A Study of the Effectives for Surface Roughness by Cutting Angle and Cutting Fluid. *Korean Soc. Manuf. Process. Eng.* **2018**, *17*, 57–62. [CrossRef]
11. Bae, J.-K.; Ahn, D.-H.; Kwon, B.-C.; Ko, S.L. Development of Efficient Brush Deburring Process for Valve Body of Auto Transmission. *Korean Soc. Manuf. Process. Eng.* **2018**, *17*, 144–152. [CrossRef]
12. Yin, C.; Heng, L.; Kim, J.S.; Kim, M.S.; Mun, S.D. Development of a New Ecological Magnetic Abrasive Tool for Finishing Bio-Wire Material. *Materials* **2019**, *12*, 714. [CrossRef] [PubMed]
13. Alam, Z.; Khan, D.A.; Jha, S. A study on the effect of polishing fluid volume in ball end magnetorheological finishing process. *Mater. Manuf. Process.* **2017**, *33*, 1197–1204. [CrossRef]
14. Liu, S.; Shan, X.; Guo, K.; Yang, Y.; Xie, T. Experimental study on titanium wire drawing with ultrasonic vibration. *Ultrasonics* **2018**, *83*, 60–67. [CrossRef] [PubMed]
15. Song, J.; Shinmura, T.; Mun, S.D.; Sun, M. Micro-Machining Characteristics in High-Speed Magnetic Abrasive Finishing for Fine Ceramic Bar. *Metals* **2020**, *10*, 464. [CrossRef]
16. Nagdeve, L.; Sidpara, A.M.; Jain, V.K.; Ramkumar, J. On the effect of relative size of magnetic particles and abrasive particles in MR fluid-based finishing process. *Mach. Sci. Technol.* **2017**, *22*, 493–506. [CrossRef]
17. Zeng, Z.; Cong, B.; Oliveira, J.; Ke, W.; Schell, N.; Peng, B.; Qi, Z.; Ge, F.; Zhang, W.; Ao, S. Wire and arc additive manufacturing of a Ni-rich NiTi shape memory alloy: Microstructure and mechanical properties. *Addit. Manuf.* **2020**, *32*, 101051. [CrossRef]

© 2020 by the authors. Licensee MDPI, Basel, Switzerland. This article is an open access article distributed under the terms and conditions of the Creative Commons Attribution (CC BY) license (http://creativecommons.org/licenses/by/4.0/).

Article

In Vitro Corrosion Behavior of Zn3Mg0.7Y Biodegradable Alloy in Simulated Body Fluid (SBF)

Cătălin Panaghie [1], Ramona Cimpoeșu [1,*], Georgeta Zegan [2,*], Ana-Maria Roman [1], Mircea Catalin Ivanescu [2], Andra Adorata Aelenei [1], Marcelin Benchea [3], Nicanor Cimpoeșu [1] and Nicoleta Ioanid [2]

1. Faculty of Materials Science and Engineering, "Gh. Asachi" Technical University from Iasi, 700050 Iasi, Romania; catalin.panaghie@student.tuiasi.ro (C.P.); ana-maria.roman@academic.tuiasi.ro (A.-M.R.); andra-adorata.aelenei@student.tuiasi.ro (A.A.A.); nicanor.cimpoesu@tuiasi.ro (N.C.)
2. Faculty of Dental Medicine, "Grigore T. Popa" University of Medicine and Pharmacy, 700115 Iasi, Romania; mircea-catalin.ivanescu@d.umfiasi.ro (M.C.I.); nicole_ioanid@yahoo.com (N.I.)
3. Faculty of Mechanical Engineering, "Gh. Asachi" Technical University from Iasi, 700050 Iasi, Romania; marcelin.benchea@tuiasi.ro
* Correspondence: ramona.cimpoesu@tuiasi.ro (R.C.); georgeta.zegan@umfiasi.ro (G.Z.)

Abstract: Biodegradable metallic materials represent a new class of biocompatible materials for medical applications based on numerous advantages. Among them, those based on zinc have a rate of degradation close to the healing period required by many clinical problems, which makes them more suitable than those based on magnesium or iron. The poor mechanical properties of Zn could be significantly improved by the addition of Mg and Y. In this research, we analyze the electro-chemical and mechanical behavior of a new alloy based on Zn3Mg0.7Y compared with pure Zn and Zn3Mg materials. Microstructure and chemical composition were investigated by electron microscopy and energy dispersive spectroscopy. The electrochemical corrosion was analyzed by linear polarization (LP), cyclic polarization (CP) and electrochemical impedance spectroscopy (EIS). For hardness and scratch resistance, a microhardness tester and a scratch module were used. Findings revealed that the mechanical properties of Zn improved through the addition of Mg and Y. Zn, Zn-Mg and Zn-Mg-Y alloys in this study showed highly active behavior in SBF with uniform corrosion. Zinc metals and their alloys with magnesium and yttrium showed a moderate degradation rate and can be considered as promising biodegradable materials for orthopedic application.

Keywords: biodegradable; Zn; corrosion; microhardness; microscratch

1. Introduction

Recently, zinc (Zn) and its alloys have attracted considerable attention and are considered promising candidates for various medical applications, due to the much more suitable degradation rate compared to magnesium (Mg) and iron (Fe) alloys. However, it is important to note that its mechanical properties need to be improved to meet the standards for medical applications. The yield strength (MPa) of Zn-based alloys presents many variations based on their added elements and states (cast, heat treated, laminated, severe plastic deformation or powder metallurgy and additive manufacturing). The values obtained experimentally vary from 50 to 500 MPa from ZnCu to ZnCuMg or ZnLi alloys, respectively [1,2]. Vickers hardness (HV) was also reported with different values from 30 to 150 HV [1].

Jain et al. [3] studied the behavior of a complex Zn alloy (96.5% Zn) in marking out a uniform corrosion and a homogeneous distribution of various reaction products obtained during the long-term immersion in SBF. Additionally, Xue et al. [4] studied a few Zn-Fe-Mg alloys in SBF and found that the Zn1Fe1Mg shows a good corrosion rate and superior mechanical properties. The corrosion rate of Zn-based alloys is influenced by the alloying elements and ranges from almost 0.050 mm/year to more than 0.300 mm/year determined by a potentiodynamic polarization test in Hank's electrolyte solution [1]. Electro-corrosion determinations show that the values of corrosion potential and the corrosion current density of zinc and Zn-based alloys are between those of magnesium- and iron-based alloys [5]. The corrosion compounds from the surface are generally made of oxides, hydroxides and phosphates of the main elements.

After years of developing strategies and methods to improve or even combat the corrosion of metallic biomaterials, there is now a growing interest in the use of corrodible metals in medical device applications. These are called biodegradable or absorbable metals [6], as they are expected to gradually corrode in vivo by generating an appropriate host response and then dissolve completely upon tissue healing [7]. A gradual transfer of the load to the healing tissue and the prevention of secondary surgery are the two main reasons why this category of metals are favorable alternatives to existing corrosion-resistant metal implants used for temporary applications [8]. Metallic materials have been investigated mainly for absorbable medical devices [9].

Zinc and zinc-based alloys are very active metals in ionic media, especially in the presence of chlorine ions. In the last decade, in the field of biodegradable metals used to make implants that after a certain time must be removed from the body, more and more attention has been paid to zinc and zinc-based alloys [10–12]. The advantages of using these materials can be summed up by the fact that they have an average corrosion rate between the corrosion rates of magnesium and the corrosion rate of iron (standard electrode potential: Mg (-2.61 V) <Zn (-1.008 V) <Fe (-0.685 V vs. ENH) [13]. Additionally, zinc is an essential element from a biological perspective [14], having an excellent hemocompatibility, and its corrosion products work as an anti-inflammatory.

Zinc has an electrode potential between those of magnesium and iron, satisfying clinical requirements, and being a good candidate as a biodegradable metal [15,16]. Moreover, zinc is widely recognized as a nutrient with an important biological function involved in synthesis, such as DNA polymerase, RNA polymerase and many transcription factors [17]. Zn was further investigated for potential clinical applications [18]. Due to the low mechanical strength and low hardness, like Mg, Zn may not meet the requirements for the mechanical properties of the implant materials. To improve their mechanical performance, a good technique is alloying with different elements. Such Zn alloys are Zn–Mg alloys [19,20], or in our case, we propose ZnMgY alloys.

This study presents results on the microstructure, mechanical properties and electro-corrosive behavior of pure Zn, Zn3Mg and Zn3Mg0.7Y alloys to determine an appropriate alloy composition for optimal medical performance.

2. Materials and Methods

Experimental materials were realized from pure zinc (99.995%) in an electrolyte bath and pure magnesium and master-alloy MgY(70/30 wt%) bought from HunanCo China, Hunan, China, molten for 600 s at 480 °C in a standard oven with induction with Argon-purged gas (~0.75 atm), Induct-Ro, Iasi, Romania. The samples consisted of three materials: pure Zn, Zn3Mg and Zn3Mg0.7Y. The materials were obtained as bars machined from ingots (approx. 110 g), which were prepared by melting from high-purity raw materials. Zn, MgY and Mg were obtained from the following material quantities: for Zn3Mg0.7Y, we used 96.0 g pure Zn, 2.9 g MgY and 1.6 pure Mg. Zinc loss by volatilization was avoided by keeping a low melting temperature and by enhancing the element dissolution in the metal bath. The samples were re-melted five times to obtain proper chemical and structural homogeneity and to reduce the voids and microcracks from the melting process.

To highlight the effect of the re-melting, we performed a non-destructive test, using fluorescent penetrant liquids, Figure 1a,b. We used the hydrophilic post-emulsification method to detect different discontinuities of the melted alloy, such as cracks or porosity. The samples were cleaned in a technical alcohol machine before testing. We used a level-four sensitivity penetrant and a hydrophilic emulsifier at 7% concentration. A non-aqueous developer was used to obtain a better contrast, which amplified the indications. Standard parameters were used as follows: penetrant dwell time: 30 min; emulsifier time: 3 min; developer time: 15 min. Inspection was performed under UV light at 3000 µW/cm^2 intensity measured at 12 inches [21]. The results of the remelting are visible primarily in the case of alloy and less of Zn pure, highlining a reduction of surface defects. Mechanical property modification was evaluated through hardness Vickers tests (HV) using HVT-1000 equipment (test force: 2.942 N–300 gF; dwell: 10 s; objective: 40× magnification, JVC TK-C92 1EC for surface image of the indentation trace) and a scratch test with CETR UMT-2 Tribometer equipment (the test consisted of the application of an increasing force of 1–15 N over a distance of 10 mm with 1 mm/s rate on the samples). Friction force (F_x) and acoustic emission (AE) data were registered during the test time, and the scratch distance was measured and recorded at a total sampling rate of 20 kHz. The apparent coefficient of friction (COF) was calculated for each sample and plotted against distance (mm). These parameters are important to establish the mechanical property modification with the addition of Mg and Y elements to pure Zn.

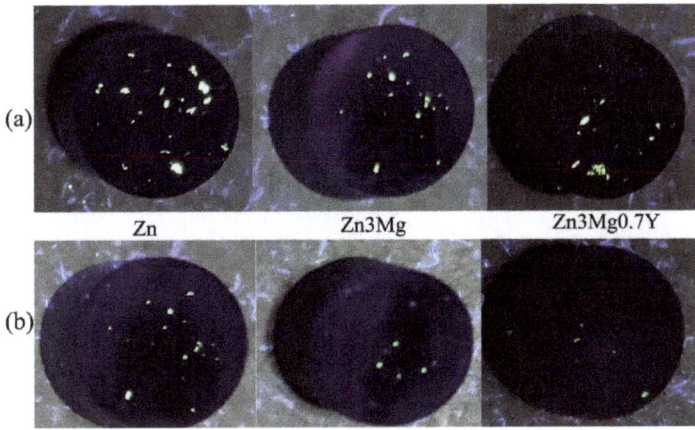

Figure 1. NDT analysis of melted Zn, Zn3Mg and Zn3Mg0.7Y alloys: (**a**) first melt sand; (**b**) after five re-melts.

Electrochemical measurements were performed with a PARSTAT 4000 electrochemical system (Princeton Applied Research, Oak Ridge, TN, USA). A C145/170 type three electrode corrosion cell (Radiometer, Neuplassans, France) was used for both the dynamic measurements and the electrochemical impedance spectroscopy determinations, which is a glass cell with the possibility of liquid corrosion; static conditions were preferred in the present measurements [22]. The placement of the samples in the working cell was performed by means of a Teflon washer with an inner diameter of 7 mm, so that, for all samples, the surface of the working electrode (the portion of the sample exposed to the corrosion environment) was also equal to 0.385 cm^2 [23]. A flat platinum electrode (S = 0.8 cm^2) was used as an auxiliary electrode, and a saturated calomel electrode as a reference. All potentials were measured in relation to this electrode, but for simplicity, this is not specified in the tables and in the text. The solution used (Simulated Body Fluid—SBF) was naturally aerated. The working conditions used in the measurements were as follows [24]:

- Linear anodic polarization for the Tafel method: potential range: (−200) ÷ (+200) mV with respect to the open circuit potential; potential scanning speed: dE/dt = 0.5 mV/s.
- Extended linear anodic polarization: potential range: (−100) ÷ (+1000) mV from the potential in open circuit; potential scanning speed: dE/dt = 1 mV/s.
- Cyclic polarization: potential range: (−500) ÷ (+2000) mV; potential scanning speed: 20 mV/s.

- Electrochemical Impedance Spectroscopy measurements, at room temperature: working potential: open circuit potential; frequency range: $10^5 \div 0.1$ Hz; potential amplitude: 10 mV.

The alloys' surfaces and microstructures were analyzed with a scanning electron microscope: Vega Tescan-LMHII, SEM, (VegaTescan, Brno—Kohoutovice, Czech Republic). Images were obtained with a Secondary Electrons (SEs) detector with 16.0 mm working distance. Determinations of chemical composition were made with Energy Dispersive Spectroscope equipment, Bruker X-Flash, Mannheim, Germany. An XRD experiment was performed with Expert PRO MPD equipment, Panalytical (XRD, Panalytical, Almelo, The Netherlands model, with a copper X-ray tube). Immersion tests were performed in SBF solution using a thermostated enclosure at 37 ± 1 °C temperature for 1, 8 and 18 days. The samples were continuously moved from a side to the other using and automatically system at each hour. The mass variation of the samples was established using a Partner analytical balance. The samples were ultrasonically cleaned in technical alcohol for 60 min after the immersion period.

3. Results

The experimental materials were investigated by chemical composition, microstructural, mechanical properties and electro-chemical behavior in a simulated body fluid electrolyte.

3.1. Chemical Composition Analysis and Microstructural Aspects

The experimental alloy Zn3Mg0.7Y was mechanically ground and polished, and chemical etching was performed in order to highlight the microstructure. The general aspects of the microstructure are given in Figure 2a. In Figure 2b, the microstructure of the Zn3Mg0.7Y alloy is presented after chemical etching. Structurally, a few different formations can be observed, and their chemical composition is presented in Table 1. The nature of the compounds is basically the same as Zn and Zn compounds (Mg2Zn11, Mg12ZnY and YZn12), as described and analyzed in a previous work through energy dispersive spectroscopy and X-ray diffraction [25].

In addition to reduced percentages of oxygen, which was removed from the results table, the main elements identified on the material were Zn, Mg and Y, as shown in Figure 2c. Quantitative results of chemical composition are given in Table 1 (mass and atomic percentages). The average chemical composition (on a 1 mm^2 area, from five different determinations) was 3.03 wt% Mg and 0.7 wt% Y. In order to establish the most important components, we performed four determinations in the areas marked in Figure 2a. The compound analyzed in point 1 is a typical YZn12 compound (formula Zn24Y2) [26]. For point 2, the matrix of the material was analyzed, which consists of a solid solution of α-Zn with dissolved Mg. Point three and four also represent ZnY and ZnMgY compounds.

Table 1. Chemical composition determination of ZnMgY alloy.

Chemical Composition	Zn		Mg		Y	
	wt%	at%	wt%	at%	wt%	at%
General (1 mm^2)	96.3	93.3	3.0	6.3	0.7	0.5
Point 1	90.0	92.4	-	-	10.0	7.6
Point 2	98.2	95.4	1.8	4.6	-	-
Point 3	91.0	93.2			8.9	6.7
Point 4	92.8	94.1	0.2	0.7	7	5.2
EDS detector Abs error [%]	2.35		1.10		1.83	

St. dev: Zn: ±1, Mg: ±0.1 and Y: ±0.05.

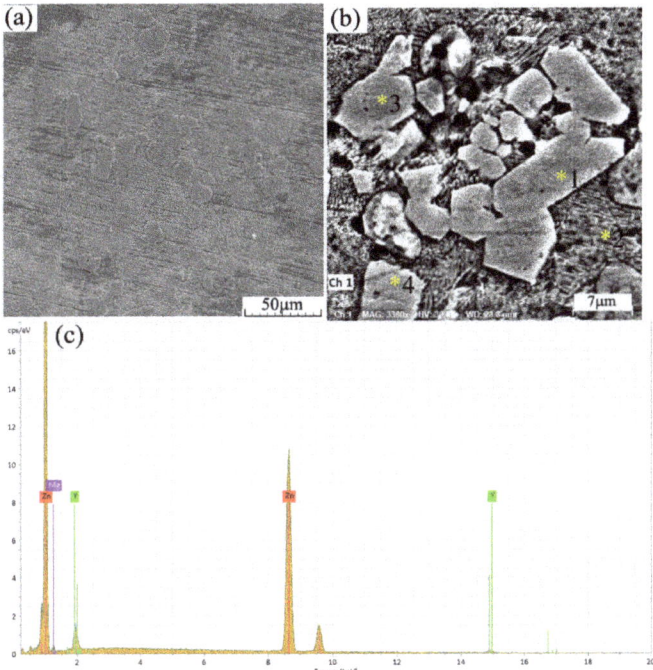

Figure 2. Structural and chemical aspects of Zn3Mg0.7Y alloy: (**a**) SEM image at 500×; (**b**) SEM image at 2500×; (**c**) energy spectrum of the chemical elements identified on area (0.063 mm^2) presented in the SEM image (**a**).

At the microscale, no structural defects, such as pores, cracks or voids, were identified after the material was re-melted five times. The experimental alloy was chemically homogeneous, without separations and agglomerations of undissolved elements.

Figure 3 shows the elemental mapping of chemical elemental components: all of them are in Figure 3a, and they are shown separately in Figure 3b–d, presenting the good chemical homogeneity of the material and highlighting the formation of ZnMg- and ZnY-based compounds. The XRD result for the Zn3Mg0.7Y alloy, as shown in Figure 3e, presented a main peak of α-Zn and compounds formed with Mg and Y. The intermetallic compound YZn12, as shown by point 1 in Table 1, was identified and confirmed on the XRD chart along with other compounds, such as MgZn2, MgZn11, ZnMg, Mg12ZnY and Mg2Zn11, as discussed and analyzed in [20].

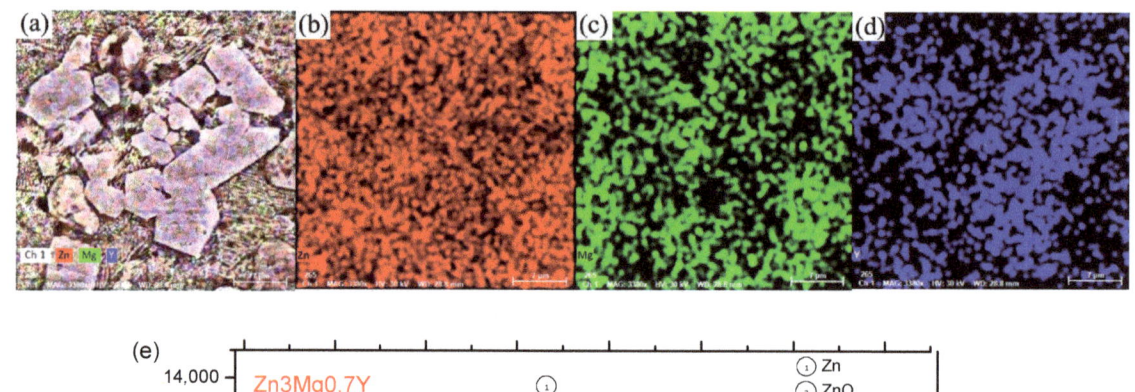

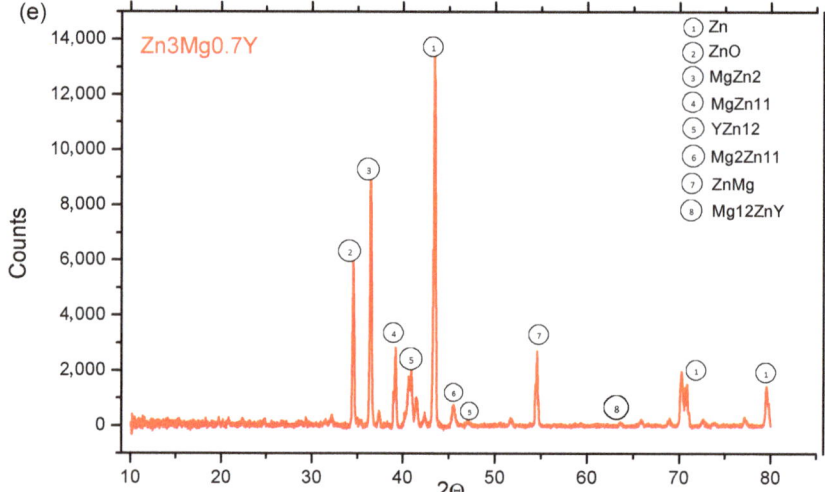

Figure 3. (**a**) Elemental mapping of all components; (**b**) Zn element; (**c**) Mg element; (**d**) Y element; (**e**) XRD spectrum of Zn3Mg0.7Y.

3.2. Microhardness and Microscratch Behavior of the Experimental Materials

The influence of the addition of Mg and Y elements on mechanical properties was obvious, first of all based on the differences between the indentor microhardness test traces, as shown in Figure 4a–d. The traces decreased in dimensions from Zn to ZnMg, and Zn3Mg0.7Y, based on the superior hardness of the compounds, formed between ZnMg and ZnY, while it did not form for pure Zn. The dimensional difference between the traces from Figure 4b,c was shown by the MgZn compounds (MgZn2, MgZn11) caught during the second test on the Zn3Mg alloy and explains the differences between the microhardness results in Table 2 (points 1 and 4).

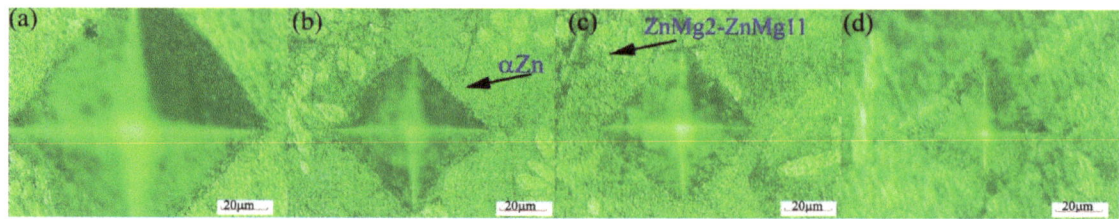

Figure 4. Optical microscopy of indented areas: (**a**) Zn; (**b**) and (**c**) ZnMg; (**d**) Zn3Mg0.7Y.

Table 2. Vickers microhardness values on cast samples.

Material	Pure Zn (HV)	Zn3Mg (HV)	Zn3Mg0.7Y (HV)
Point 1	45.1	118.0	118.6
Point 2	50.7	121.0	126.7
Point 3	51.0	129.3	133.8
Point 4	48.0	127.0	120.1
Point 5	62.5	125.7	132.4
Average	51.5	124.2	126.3

The formation of ZnMg and ZnY compounds improved the microhardness of the pure Zn by more than two-fold. The contribution of Y and especially the YZn compounds was the increase in the microhardness of the material. Except two areas—probably with a lower content of ZnMg and ZnY compounds, points 1 and 4—all the results for Zn3Mg0.7Y material presented a higher hardness compared to the Zn3Mg alloy.

Li et al. [27] obtained (microalloyed with Al, Mn, Cu and Ag) a Vickers hardness of 51 ± 3.4 HV for Zn, and after alloying with Li, a value of 90 ± 6.9 HV. Yang et al. [28] also present an increase in the Vickers hardness of pure zinc with the addition of Ca and Cu from 32.12 HV to 71.83 HV Zn-1Ca-0.5Cu. Xivei Liu obtain the following values of microhardness: 93.71 HV for the Zn–1Mg–0.1Sr alloy, and 109.34 HV for the Zn–1Mg–0.5Sr alloy, which are higher than those of pure Zn, suggesting the effectiveness of alloying in improving its mechanical property [29]. Pachla et al. present Vickers hardness values for samples Zn0.5Mg, Zn1Mg and Zn1.5Mg after a hot extrusion process: 75 HV, 95HV and 115 HV, respectively [30,31].

The scratch test is regularly used for the assessment of the cohesive and adhesive strength of thin films and coatings. By default, its evaluation is based on the analysis of the depth–load–time record and the microscopic observation of residual scratch grooves [32]. In our case, we tested the materials in order to compare their behavior for similar reasons. In Figure 5, the scratch behavior of the samples is presented for Figure 5a F_x vs. Y (mm), Figure 5b COF vs. Y (mm) and Figure 5c AE vs. Y (mm). The quantitative results are given in Table 3. Both the force F and friction coefficient COF presented similar behaviors with higher values for pure Zn and similar variations for Zn3Mg and Zn3Mg0.7Y. Based on the behavior of pure Zn, the increase in F_x and COF values can be attributed to overlapping of the soft Zn matrix, especially on the first 4 µm of the scratch.

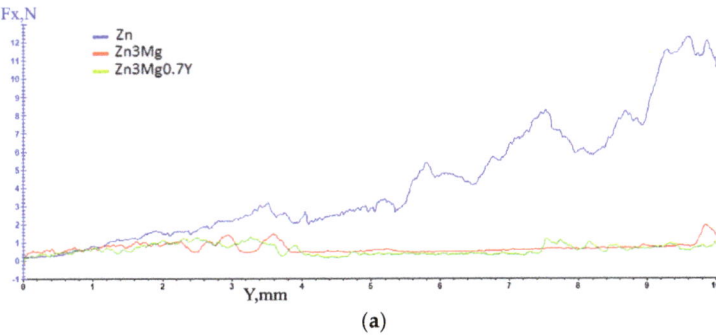

(a)

Figure 5. Cont.

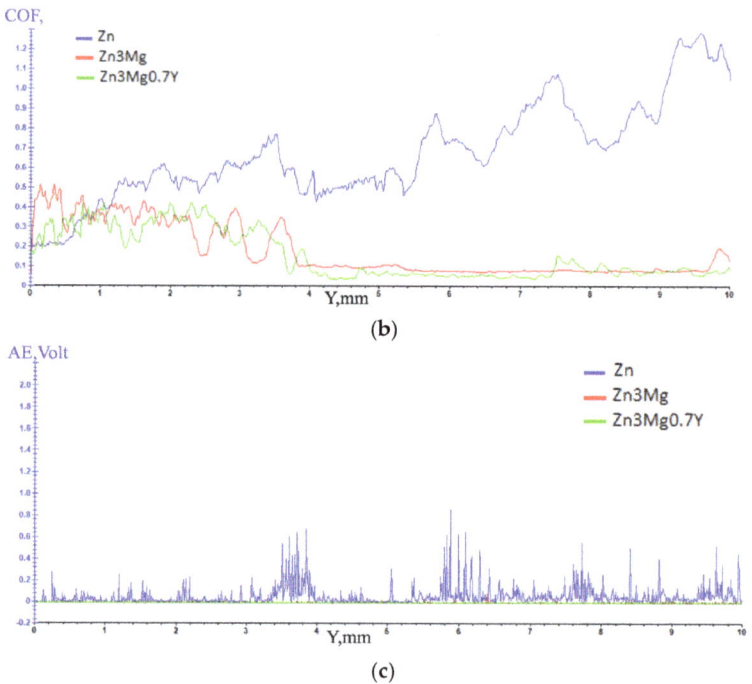

Figure 5. Scratch behavior of the samples: (a) F_x vs. Y (mm); (b) COF vs. Y (mm); (c) AE vs. Y (mm).

Table 3. Scratch resistance test results.

Material	Average Fx (N)	Average AE (Volt)	Average COF
Zn	4.230	0.052	0.674
Zn3Mg	0.695	0.005	0.177
Zn3Mg0.7Y	0.595	0.004	0.156

The visual analysis of the residual groove provides the most detailed description of the final damage of the surface (crack patterns, extent of plastic deformation, delamination, etc.), but it may be a time-consuming approach. Although the continuous recording of indenter penetration depth and applied load offers instantaneous information about the performance of the tested material, it may not provide a sufficient description of the sample's deformation behavior [33]. Therefore, other complementary techniques for the description of the deformation response to scratch loading are desirable. The continuous recording of acoustic emissions (AE) generated during the test could be a possible solution. Especially the ability of the AE method to detect the very first and even subsurface failures of the material is of the utmost importance and otherwise inaccessible by standardly used techniques [34,35].

Figure 6 shows given SEM scratch stain images in Figure 6a pure Zn, Figure 6b Zn3Mg and Figure 6c Zn3Mg0.7Y. Different widths of the scratch stain were observed with a dimension three times bigger for pure Zn (~400 µm) compared to ZnMg (~130 µm) and ZnMgY (~125 µm). Overlapped material is observed in case of pure Zn and no scratch or voids are present at the edges of the scratch stain.

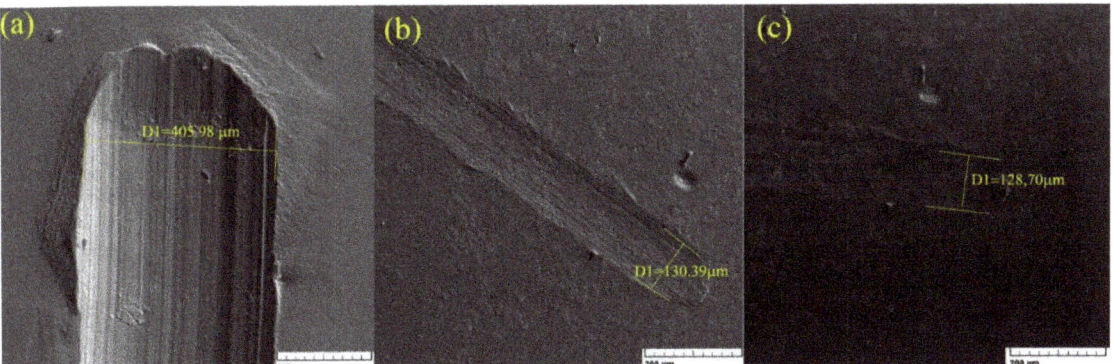

Figure 6. SEM scratch stain: (**a**) pure Zn; (**b**) Zn3Mg; (**c**) Zn3Mg0.7Y.

The soft nature of pure Zn increased the value of F_x (N) force necessary to scratch the material, and a higher value of the friction coefficient was obtained for pure zinc. The friction coefficient as 3.8 times higher for pure Zn compared with Zn3Mg and 4.32 times higher than that of the Zn3Mg0.7Y alloy.

The AE (acoustic emission) values were appropriate for Zn3Mg and Zn3Mg0.7Y alloys and at a low intensity compared to pure zinc, which presented high variations, as shown in Figure 5c, on different areas. The AE values were ten times smaller than those of pure Zn, which is in accordance with the SEM images of the scratch stains and variations of COF and F_x.

3.3. Electro-Corrosion Behavior of Zn, ZnMg and ZnMgY Materials in SBF Electrolyte

The corrosion potential, $E_{corr} = E\ (I = 0)$, is a measure of the corrosion tendency of a metal or alloy immersed in a given electrolytic medium (thermodynamic probability of corrosion). In fact, this is the potential value (measured in relation to the reference electrode—in this case, the saturated calomel electrode) where the anodic and cathodic reactions rates meet. Very high negative corrosion potential values indicate a very high tendency for zinc and zinc-based alloys to corrode. In the case of pure zinc, the increase in the negative value of the corrosion potential indicated a slight increase in the tendency of corrosion, probably caused by the increase in the surface of the sample due to corrosion (the surface was no longer flat but rough).

In the case of the Zn3Mg0.7Y alloy, the evolution of the corrosion potential indicated a slight tendency of passivation (very small), while for the Zn-Mg alloy, the variation was more complex.

The corrosion current and, directly related to it, the corrosion rate had different evolutions for the three alloys depending on the immersion time in SBF. Thus, in the case of pure zinc, the reaction rate increased appreciably with the storage time, which was also caused by the increase in the surface roughness corresponding to generalized corrosion (Figure 7). When calculating the corrosion rate, the initial plan surface was taken. The parameters of the instantaneous corrosion process evaluated by the Tafel method are presented in Table 4.

Table 4. Instantaneous corrosion process parameters according to the immersion time of the alloys in SBF.

Parameter	Zn			Zn-3Mg			Zn3Mg0.7Y		
	0 Days	8 Days	18 Days	0 Days	8 Days	18 Days	0 Days	8 Days	18 Days
$E(I = 0)$, mV	−1078	−1.081	−1105	−1060	−1079	−1050	−1106	−1007	−1086
j_{cor}, μA/cm^2	11	25	49.14	31.57	35.07	12.46	18.69	36.87	17.73
v_{corr}, mm/Y	0.2940	0.6427	1.3123	0.8432	0.9229	0.3329	0.4993	0.9847	0.4737
β_a, mV/decade	86.43	56.02	74.47	55.64	53.79	49.44	80.61	67.37	83.73
β_c, mV/decade	322.19	185.09	136.97	222.22	177.34	251.21	132.77	221.15	110.4

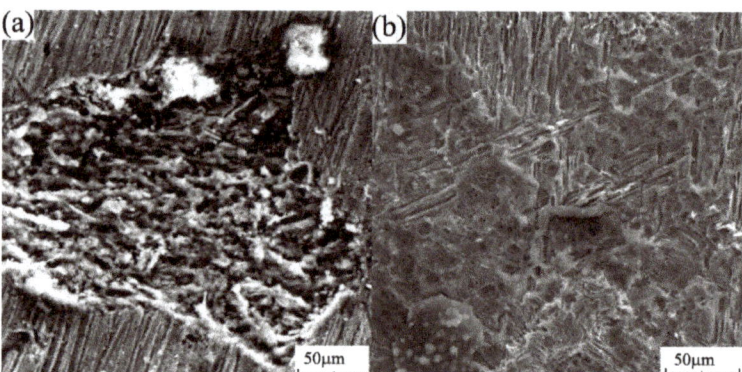

Figure 7. The roughness evolution for the zinc sample after maintenance in SBF (500×). (**a**) generalized corrosion; (**b**) solid reaction products.

In the case of the Zn-Mg alloy, the corrosion rate was still high in the initial moments, this being probably due to the much higher reactivity of magnesium than that of zinc. After 8 days of immersion, the reaction rate increased, which is unlikely due to the roughness, but after 18 days of immersion, it decreased appreciably, reaching a value very close to the corrosion rate of pure zinc. This may be due to the prolonged immersion in the solution of this initial alloy in the solubilization of magnesium until total depletion on the surface and the subsequent corrosion of zinc. The deposition on the surface of the sample of some solid reaction products can be added to this, as in the case of zinc (Figure 7).

In Figure 8, for the sample maintained for 18 days in SBF, the cavities from which the magnesium dissolved can be observed.

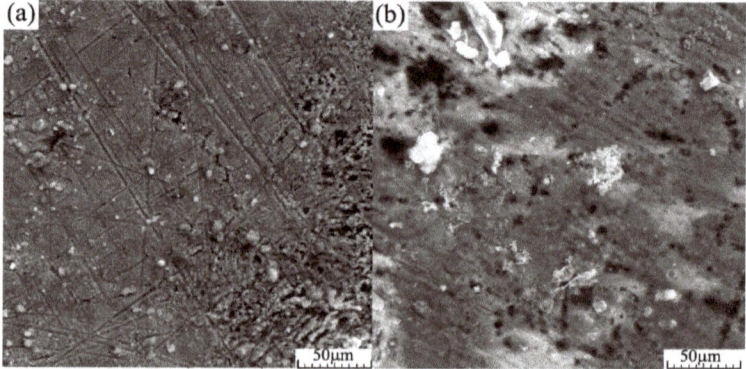

Figure 8. The evolution of the Zn-Mg sample surface after maintenance in SBF (500×). (**a**) cavities; (**b**) solid reaction products.

In the case of the Zn3Mg0.7Y alloy test, the corrosion rate increased in the initial moments and after 18 days decreased appreciably. This behavior is due to the formation of a crust from insoluble solid reaction products, as can be seen in Figure 9.

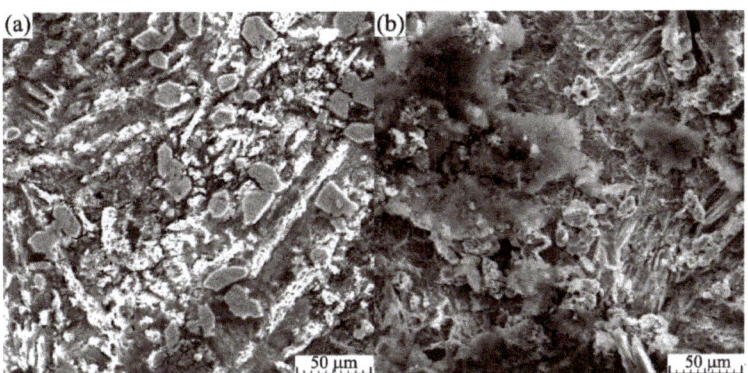

Figure 9. The surface evolution of the alloy sample after maintaining in SBF (500×). (**a**) insoluble solid reaction products; (**b**) solid reaction products.

Tafel slopes provide information on the reaction mechanism. In this case, the low value of the anodic slope indicated that the anodic reaction ($Zn \rightarrow Zn^{2+} + 2e^-$) is the active reaction. As suggested by the higher cathodic slope, the corrosion process is under concentration polarization control. Activation control is determined by the rate of electron transfer from the anode to the cathode. In the presence of dissolved oxygen at the cathode, the following reaction takes place: $1/2\ O_2 + H_2O + 2e^- \rightarrow 2OH^-$, followed by the reaction $Zn^{2+} + 2OH^- = Zn(OH)_2$ [36].

All the linear polarization curves recorded for the three alloys after various storage periods in SBF had the general appearance shown in Figure 10a.

The curve has two distinct segments: a nonlinear segment at very low currents, denoted as the mixed potential domain, located around the corrosion potential, and a linear portion, starting from a threshold potential, E_{gc}, corresponding to generalized corrosion. For potentials greater than E_{gc}, the corrosion current increased in direct proportion to the overpotential applied to the metal and can be expressed by the following equation:

$$I(mA) = a.E(mV) + b$$

Unfortunately, for these systems, the polarization curves and, thus, the Evans diagram were very different from the classical curves, in which the two branches (anodic and cathodic) were symmetrical. Due to this, the values of the Tafel slopes highly depended on the way the data were processed, see Figure 10b, especially the size of the potential range around the chosen corrosion potential. The linear potentiometry parameters obtained are presented in Table 5. The corrosion current (i_{corr}) values increase along with the increase from 2.28 µA, for Zn to 15.57 µA for Zn3Mg.7Y. The corrosion rate of Zn3Mg0.7Y is 83.38 mpy, higher than Zn and Zn3Mg. This is because of the non-homogeneous structure caused by the formation of new compounds with Y. The values of the constants a and b for the alloys studied as a function of the immersion time in the solution are presented in Table 6. In the last column of Table 6, the linear correlation coefficients for the respective straight sections are presented. The corrosion potential presented was evaluated by the Tafel method, from the linear polarization curves recorded at a sweep speed potential of 1 mV/s. Their values were comparable to those in Table 4, obtained by the same method from the curves recorded at a scan rate potential of 0.5 mV/s, being only slightly higher, but the increases were not significant.Unfortunately, for these systems, the polarization curves and, thus, the Evans diagram were very different from the classical curves, in which the two branches (anodic and cathodic) were symmetrical. Due to this, the values of the Tafel slopes highly depended on the way the data were processed, see Figure 10b, especially the size of the potential range around the chosen corrosion potential. The linear potentiometry parameters obtained are presented in Table 5. The corrosion current (i_{corr}) values increase along with the increase from 2.28 µA, for Zn to 15.57 µA for Zn3Mg.7Y. The corrosion rate of Zn3Mg0.7Y is 83.38 mpy, higher than Zn and Zn3Mg. This is because of the non-homogeneous structure caused by the formation of new compounds with Y.

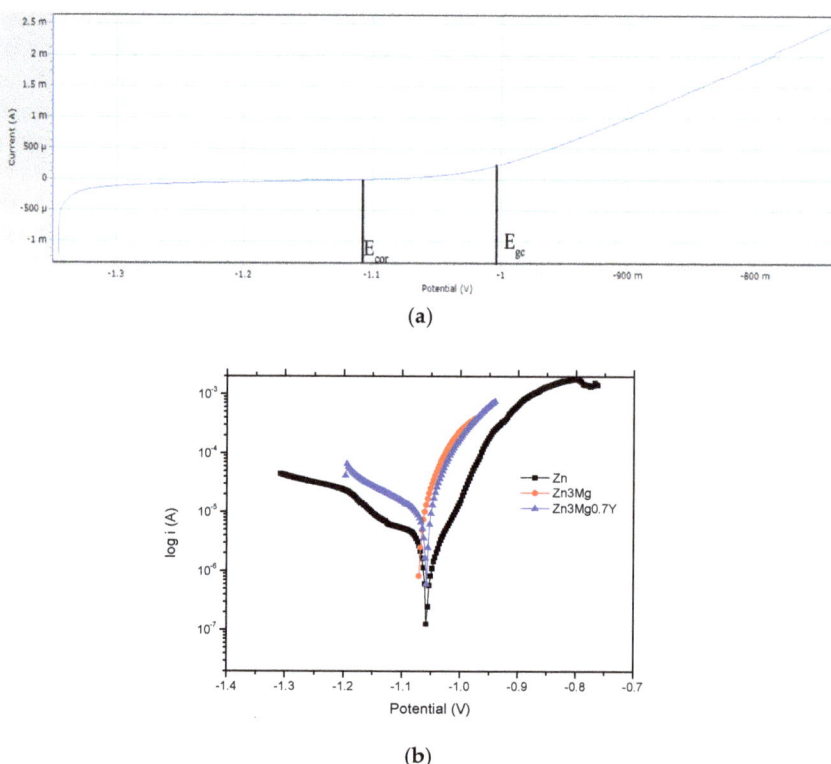

Figure 10. (a) The linear polarization curve for Zn with freshly ground surface in SBF, (b) Tafel slopes.

Table 5. Linear potentiometry parameters.

Sample	E_{cor} (V)	i_{corr} (µA)	V_{corr} (mpy)
Zn	1.05	2.28	8.58
Zn3Mg	1.07	13.32	50.05
Zn3Mg0.7Y	1.13	15.57	83.38

An order of variation of the corrosion potential could not be established either in the case of the same alloy or between the alloys, the differences being located in the limits of the experimental errors. The main metal in the alloy, in very large quantities, was zinc, and an average value of the average corrosion potential was determined: $(E_{cor})_{average}$ = −1114 mV for dE/dt = 1 mV/s and $(E_{cor})_{average}$ = −1086 mV for dE/dT = 0.5 mV/s.

Table 6. Dependence of the corrosion current on the overcurrent applied to the alloy, at potentials higher than E_{gc}.

Alloy	Time	E_{cor} (mV)	E_{gc} (mV)	I(mA) = a.E(mV) + b		r_L
				a (mA/mv)	b (mA)	
Zn	0 days	−1121	−950	0.008346	8.10571	0.997
	8 days	−1108	−1000	0.009013	9.11235	0.999
	18 days	−1136	−1000	0.010086	10.4150	0.999
Zn3Mg	0 days	1108	−950	0.014225	13.91910	1.000
	8 days	−1101	−990	0.009951	9.89128	1.000
	18 days	−1092	−995	0.007800	6.95717	1.000
Zn3Mg0.7Y	0 days	−1106	−1000	0.015565	14.86469	1.000
	8 days	−1116	−990	0.008197	8.09233	1.000
	18 days	−1137	−1000	0.009576	9.24229	0.997

The slope of the lines describing the influence of the overcurrent applied to the metal on the corrosion current, and implicitly on the corrosion rate, in the case of zinc increased with the increase in the intercept with the potential axis (b), which translated into a slight increase in the general corrosion process.

In the case of the Zn3Mg alloy, both the slope and the original cut decreased with the immersion time in the solution, thus marking a decrease in the corrosion rate over time. This is explained by the fact that in the initial period, magnesium dissolves first, which is the most electronegative metal.

In the case of the Zn3Mg0.7Y, there was no orderly variation for either the slope or the intercept at the origin. The cyclic voltammograms for the three alloys for various immersion intervals in SBF all had the same force as that shown in Figure 11.

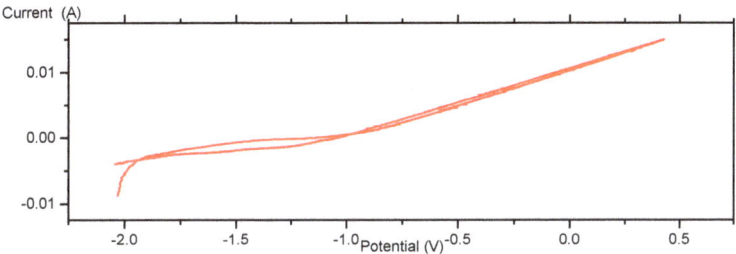

Figure 11. Cyclic voltammogram for zinc with freshly ground surface in SBF; dE/dt = 10 mV/s.

For all curves, the linear current–voltage dependence started from a voltage value of −1000 mV. It was noticed that the return branch (cathodic curve) overlapped almost perfectly on the direct branch (anodic curve). This means that the generalized corrosion maintained by a large voltage value did not appreciably alter the active surface, nor did it produce passivation phenomena by the deposition of reaction products.

The electrochemical impedance spectroscopy data were processed with the SZSimp-Win software, which uses the least nonlinear squares method to obtain the most appropriate values of the equivalent circuit elements tested. The parameters that best describe (fit) an equivalent circuit were obtained by minimizing the function χ^2, defined as the sum of the squares of the residuals (the differences between the calculated values and the experimental values):

$$\chi^2 = \sum_{i=0}^{n} \left[W_i' \left(Z_i' \left(\omega^i, \vec{p} \right) - a_i \right)^2 + W_i'' \left(Z_i'' \left(\omega_i, \vec{p} \right) - b_i \right)^2 \right] \quad (1)$$

where n is the number of points and W_i is the weighting coefficients. A value of χ^2 equal to 10^{-4} translates into a relative error of the measured current of 0.01, i.e., 1%.

For a certain equivalent circuit to be suitable for describing the physical condition of the alloy surface, the minimum value of χ^2 is not sufficient, but the errors associated with each circuit element must be below 5%.

Taking into account these conditions, for the optimal fit of the experimental data, the circuits shown in Figure 12 were established.

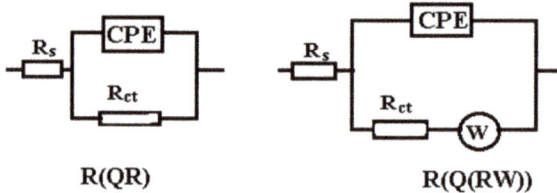

Figure 12. Equivalent circuits used for processing EIS data on the three alloys depending on the storage time.

The R (QR) circuit is suitable for describing a system in which a single reaction takes place on the surface of the alloy and corrosion is controlled entirely by the transfer of charges through the double-electric layer. In this circuit, R_s is the resistance of the solution between the electrode surface and the reference electrode, R_{ct} is the opposite resistance to the charge transfer and CPE is a constant phase element introduced instead of the capacity of the double-electric layer (C_{dl}) for a better adjustment. This showed good experimental data. The introduction of this element was necessary due to the fact that the surface of the working electrode is not homogeneous and the electrical capacity is frequency dependent.

If for a capacitor the impedance is equal to $Z_C = 1/(j\omega C)$, in the case of the constant phase element, the impedance is evaluated in accordance with [6–9]:

$$Z_{CPE} = \frac{1}{Q(j\omega)^n} \quad (2)$$

where Q is a constant proportional to the active area (area exposed to corrosion), $<Q> = \Omega^{-1}s^n/cm^2 \equiv S.s^n/cm^2$, ω is the angular frequency ($\omega = 2\pi f$; f = frequency of the applied alternating current), j is the imaginary number and $j = (-1)^{1/2}$. A consequence of this simple relationship is that the phase angle of the CPE is independent of frequency and has a value of $(90°)^n$, which is also the reason that it is called a constant phase element. The values of the circuit elements for this equivalent circuit are shown in Table 7. The Nquist and Bode diagrams for Zn, ZnMg and ZnMMgY obtained are presented in Figure 13a–c.

Table 7. Circuit R(QR) element values.

Alloy	Immersion Time	$10^3 \cdot \chi^2$	R_s Ohm.cm^2	$10^4 \cdot Q$ S.s^n/cm^2	n	R_{ct} Ohm.cm^2
Zn	0 days	2.40	27.89	0.371	0.705	920.4
	8 days	1.52	47.93	2.660	0.663	269.6
	18 days	0.56	36.47	1.539	0.554	218.2
Zn3Mg	0 days	5.09	27.89	0.1245	0.597	358.4
	8 days	0.80	45.36	30.170	0.379	171.9
	18 days	-	-	-	-	-
Zn3Mg0.7Y	0 days	0.76	25.93	0.4056	0.614	535.1
	8 days	-	-	-	-	-
	18 days	7.28	27.71	5.987	0.515	220.3

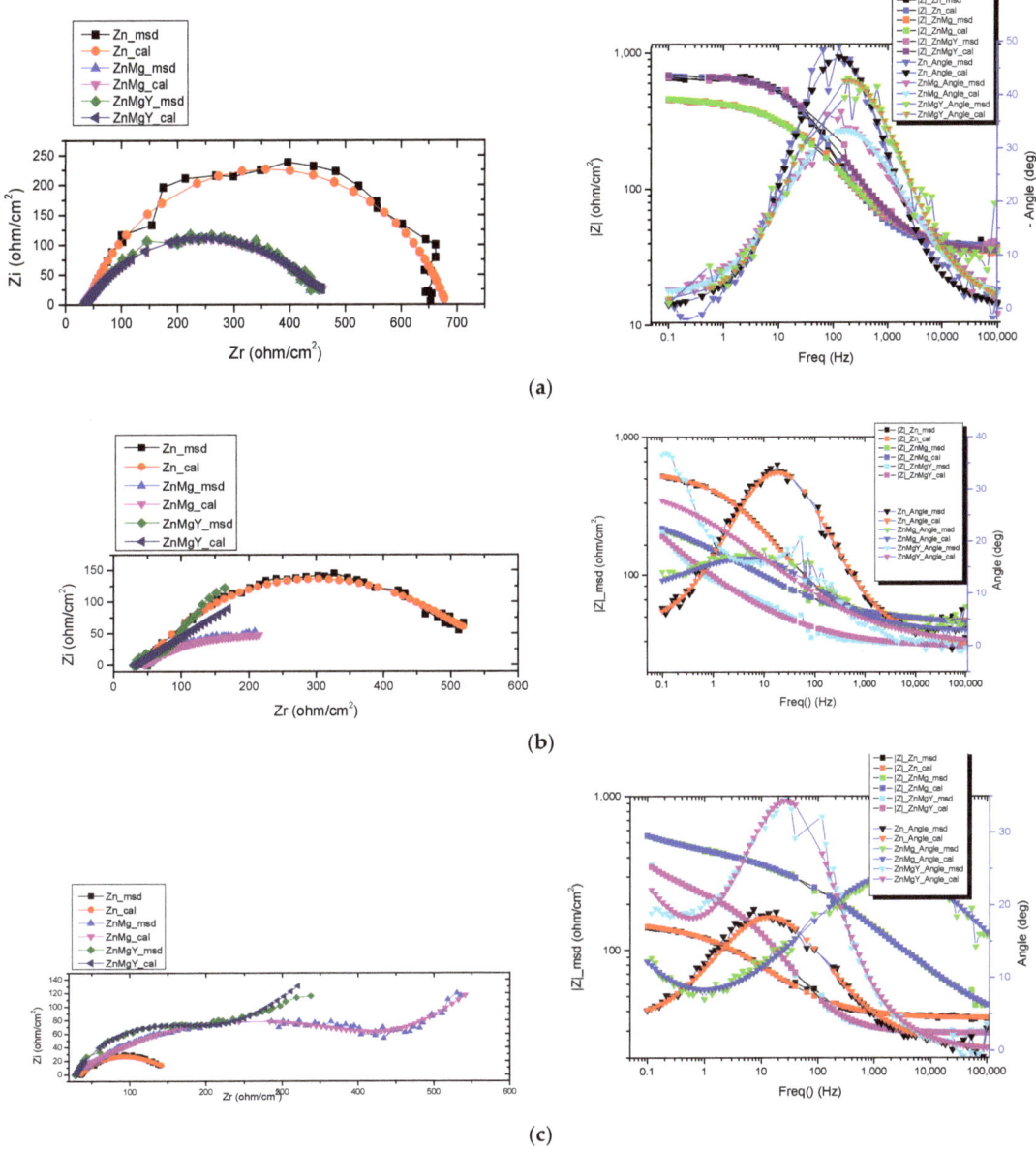

Figure 13. Nquist and Bode diagrams for Zn, ZnMg and ZnMMgY at: (**a**) 0 days; (**b**) 8 days; (**c**) 18 days.

This circuit satisfactorily describes the experimental data for zinc, and for the freshly ground surface and for samples immersed in SBF 8 or 18 days. The percentage errors for the circuit elements were approximately 1% and even lower, except for *Q*, where they were of the order of 3–4%, but located in the reliable range. It should be noted that in the case of zinc, the load transfer resistance (R_{ct}) decreased appreciably with the immersion time in the solution, which increased the reaction rate, resulting in very good agreement with that found for the instantaneous corrosion rate evaluated in the linear polarization curve (Tafel method).

The exponent n, which gives an indication of the deviation from the ideality of the capacity of the double-electric layer, which also decreases with the immersion time, a decrease attributed to the increase in the degree of surface roughness due to corrosion was observed. This behavior also indicates that, although it is possible that insoluble reaction products, such as ZnO or ($Zn_5(OH)_8Cl_2 \cdot 2H_2O$) (simonkolleit) [25], may form during storage in solution, they are probably porous and do not act as a barrier to the reaction.

In the case of the Zn-Mg alloy, the R (QR) circuit satisfactorily described the experimental data only for the freshly ground sample and for the sample maintained for 8 days in SBF, but here too the polarization resistance decreased and the corrosion rate increased. Moreover, the frequency exponent in the expression of the constant phase element showed unexpectedly low values, probably marking an appreciable deterioration of the alloy surface and maybe some local deposits.

In the case of Zn3Mg0.7Y, only the equivalent circuit could not be used for the test held for 8 days in solution. Analyzing the evolution of the constant Q as a function of immersion time, this varied randomly, with very large oscillations, both between different alloys and for the same alloy at different immersion times, this circuit element being more sensitive to experimental errors, exemplified by the percentage errors, between 3 and 5%, obtained when fitting the curve.

For the sample maintained for 18 days in SBF, the frequency exponent was very close to a value of 0.5, indicating the possibility of a diffusion phenomenon. As in the case of the other two alloys, some values close to 0.5 were encountered; we tried to use the second equivalent circuit, as shown in Figure 12, which contained a diffusion impedance. The values of the circuit elements, evaluated on the basis of the same experimental data, are presented in Table 8.

Table 8. Values of circuit elements R (Q (RW)).

Alloy	Immersion Time	$10^3 \chi^2$	R_s Ohm.cm^2	$10^3 Q$ S.s^n/cm^2	n	R_{ct} Ohm.cm^2	W S.s$^{1/2}$/cm^2
Zn	0 days	1.52	28.22	3.323	0.718	867.9	0.008637
	8 days	1.18	47.58	3.004	0.643	447.1	0.06193
	18 days	1.15	35.92	1.876	0.531	205.9	0.04964
Zn3Mg	0 days	0.90	29.51	0.0710	0.655	291.4	0.05335
	8 days	0.89	47.19	0.3250	0.636	503.8	0.065730
	18 days	0.68	25.77	0.1659	0.420	455.2	0.008236
Zn3Mg0.7Y	0 days	0.62	26.18	0.03832	0.620	524.1	0.03609
	8 days	1.35	32.46	0.2468	8.837	28.75	0.008623

Comparing these values with the values of the parameters for the R (QR) circuit, it was found that they were very close, which is additional proof that these circuits also describe the same state.

The need to introduce a diffusion impedance, even in the absence of a film adsorbed or adhering to the surface of the metal, may seem at least risky. This can be explained by considering the existence of a local diffusion in a nanometer-sized film in the reaction zone on the metal surface [37]. Zinc is a very active metal in ionic media containing chlorine, with its surface suffering a generalized, uniform corrosion. Corrosion occurred in a single reaction, and the reaction products were soluble. There was a uniform concentration of ions and electrons in a nanometer-sized layer on the surface. The mobility of electrons is much higher than that of ions [38]. As the electrons became free in the system, the electric field relaxed rapidly, which led to a local charge neutrality, and the transport of the charge carriers was limited by diffusion through this overloaded nano-layer with charges. The values of the parameter χ^2 indicated that the last two circuits (one is enough) described the experimental data much better, followed by those for the R circuit (QR).

In all cases, the immersion tests present a decrease in samples mass with corrosion compounds that pass from the material surface to electrolyte solution, values are given in Table 9. Using the follow densities: [g/cm^2]: 7.13 for Zn, 6.52 for ZnMg and 6.45 for Zn3Mg0.7Y we calculate the corrosion rate of the material in SBF electrolyte based on formula: CR = CR = (8.76*10,000*mass loss)/(Total area*time*density) [mm/year] [39]. The differences between the cleaned and uncleaned sample are given by the instable compounds formed on the surface during immersion and that pass to solution after ultrasonication of the samples.

Table 9. Mass loss after 18 days of immersion in SBF electrolyte.

Material/Alloy	Initial Mass [g]	After Immersion (18 Days) [g]	After Ultrasound Cleaning [g]	Loss Mass During Immersion [g]	Corrosion Rate (CR) [mm/Year]
Zn	3.7153	3.7012	3.7001	0.0011	0.13
Zn3Mg	4.2286	4.2104	4.2095	0.0009	0.23
Zn3Mg0.7Y	2.5994	2.5883	2.5879	0.0004	0.12

The macro and micro aspects of the compounds passed from the alloy surface into the electrolytic solution are shown in Figures 14a and 14b, respectively. Generally, small and large parts of the material can be observed. The larger parts are usually agglomerations of small round oxides, Figure 14b. At microscale, the minimum diameter measured using VegaTC software was around 75 µm and the maximum one at 1600 µm, with an average of 480 µm. At micro scale (analyze of the parts from Figure 14b) the minimum value measured was of 2 µm and the maximum of 5.07 µm and an average value of 3.29 µm (50 determinations) and a standard deviation of ±0.74.

Figure 14. SEM images of compounds passed from Zn3Mg0.7Y in electrolyte solution after 18 days 25× in (**a**) and 1000× in (**b**).

As marked in Figure 14a, the chemical composition on different areas was analyzed and the quantitative results are given in Table 10.

All the products present a high amount of oxygen, mainly oxides passing from the material surface to electrolyte solution, and also chlorine and carbon (due to the formation of carbonates) based on the compounds identified.

Table 10. Chemical composition of the compounds from electrolyte solution.

Zn3Mg0.7Y Alloy	Zn		Mg		Y		O		C		Cl	
	wt%	at%	wt%	at%	wt%	at%	wt%	at%	wt%	at%	wt%	at%
Area 1	58.3	24.6	0.7	0.8	0.9	0.3	24.6	42.39	13.1	30.1	2.4	1.9
Area 2	50.3	19.9	0.5	0.6	0.4	0.1	29.4	47.61	12.3	26.6	7.1	5.2
Area 3	55.8	22.1	0.8	0.8	0.5	0.1	25.6	41.44	15.9	34.4	1.4	1.1
EDS detector error %	1.5		0.3		0.1		1.0		0.2		0.1	

C percentage values are strongly influenced by the double layered carbon tape used to fix the corrosion compounds for the SEM-EDS analysis.

4. Conclusions

The article presents the experimental results of a new alloy, Zn3Mg0.7Y, with possible applications in the field of biodegradable metallic elements. The conclusions can be summarized as follows:

- A new alloy, ZnMgY, with a good structural and chemical homogeneity, was obtained using an induction furnace;
- After five re-melting stages, no pores, voids or microscratches were observed through the penetrant liquid NDT method;
- The main compounds of Zn3Mg0.7Y were determined, and their influence on mechanical properties compared to pure Zn and Zn3Mg alloys was evaluated;
- An increase in microhardness was obvious with the addition of Mg and Y elements;
- Fx and COF of the pure Zn were decreased with the addition of Mg and Y.

For all the curves, the linear current–voltage dependence started from a voltage value of −1000 mV. It was found that the return branch overlapped almost perfectly on the direct branch (anodic curve). This means that the generalized corrosion maintained by large voltage values did not appreciably alter the active surface, nor did it produce passivation phenomena by the deposition of reaction products.

Author Contributions: Conceptualization, C.P., R.C. and N.C.; Formal analysis, G.Z., A.-M.R., A.A.A., M.B. and N.I.; Funding acquisition, N.C.; Investigation, C.P., R.C., A.-M.R., M.C.I., A.A.A. and M.B.; Methodology, C.P., R.C., G.Z., A.-M.R., M.C.I. and N.I.; Project administration, N.C.; Resources, G.Z.; Software, M.B.; Supervision, N.C.; Validation, G.Z., A.-M.R., M.C.I. and N.I.; Visualization, N.I.; Writing—original draft, R.C. and N.C.; Writing—review and editing, A.-M.R. All authors have read and agreed to the published version of the manuscript.

Funding: This research was funded by publications grant of the TUIASI, project number GI/P2/2021.

Institutional Review Board Statement: Not applicable.

Informed Consent Statement: Not applicable.

Conflicts of Interest: The authors declare no conflict of interest.

References

1. Shi, Z.-Z.; Gao, X.-X.; Zhang, H.-J.; Liu, X.-F.; Li, H.-Y.; Zhou, C.; Yin, Y.-X.; Wang, L.-N. Design biodegradable Zn alloys: Second phases and their significant influences on alloy properties. *Bioact. Mater.* **2020**, *5*, 210–218. [CrossRef] [PubMed]
2. Liu, H.; Huang, H.; Zhang, Y.; Xu, Y.; Wang, C.; Sun, J.; Jiang, J.; Ma, A.; Xue, F.; Bai, J. Evolution of Mg–Zn second phases during ECAP at different processing temperatures and its impact on mechanical properties of Zn-1.6Mg (wt.%) alloys. *J. Alloys Compd.* **2019**, *811*, 151987. [CrossRef]
3. Jain, D.; Pareek, S.; Agarwala, A.; Shrivastava, R.; Sassi, W.; Parida, S.K.; Behera, D. Effect of exposure time on corrosion behavior of zinc-alloy in simulated body fluid solution: Electrochemical and surface investigation. *J. Mater. Res. Technol.* **2021**, *10*, 738–751. [CrossRef]
4. Xue, P.; Ma, M.; Li, Y.; Li, X.; Yuan, J.; Shi, G.; Wang, K.; Zhang, K. Microstructure, Mechanical Properties, and In Vitro Corrosion Behavior of Biodegradable Zn-1Fe-xMg Alloy. *Materials* **2020**, *13*, 4835. [CrossRef]
5. Dong, H.; Lin, F.; Boccaccini, A.R.; Virtanen, S. Corrosion behavior of biodegradable metals in two different simulated physiological solutions: Comparison of Mg, Zn and Fe. *Corros. Sci.* **2021**, *182*, 109278. [CrossRef]

6. Hermawan, H. Updates on the research and development of absorbable metals for biomedical applications. *Prog. Biomater.* **2018**, 7, 93–110. [CrossRef]
7. Witte, F.; Eliezer, A. Biodegradable Metals. In *Degradation of Implant Materials*; Eliaz, N., Ed.; Springer: New York, NY, USA, 2012; pp. 93–109.
8. Pietrzak, W.S.; Eppley, B.L. Resorbable polymer fixation for craniomaxillofacial surgery: Development and engineering paradigms. *J. Craniofac. Surg.* **2000**, 11, 575–585. [CrossRef]
9. Im, S.H.; Jung, Y.; Kim, S.H. Current status and future direction of biodegraable metallic and polymeric vascular scaffolds for next-generation stents. *Acta Biomater.* **2017**, 60, 3–22. [CrossRef]
10. Bowen, P.K.; Drelich, J.; Goldman, J. Zinc exhibits ideal physiological corrosion behavior for bioabsorbable stents. *Adv. Mater.* **2013**, 25, 2577–2582. [CrossRef]
11. Purnama, A.; Hermawan, H.; Mantovani, D. Biodegradable metal stents: A focused review on materials and clinical studies. *J. Biomater. Tissue Eng.* **2014**, 4, 868–874. [CrossRef]
12. Li, H.; Zheng, Y.; Qin, L. Progress of biodegradable metals. *Prog. Nat. Sci. Mater. Int.* **2014**, 24, 414–422. [CrossRef]
13. Cheng, J.; Liu, B.; Wu, Y.H.; Zheng, Y.F. Comparative in vitro study on pure metals (Fe, Mn, Mg, Zn and W) as biodegradable metals. *J. Mater. Sci. Technol.* **2013**, 29, 619–627. [CrossRef]
14. Frederickson, C.J.; Koh, J.Y.; Bush, A.I. The neurobiology of zinc in health and disease. *Nat. Rev. Neurosci.* **2005**, 6, 449–462. [CrossRef] [PubMed]
15. Zheng, Y.F.; Gu, X.N.; Witte, F. Biodegradable metals. *Mater. Sci. Eng. R* **2014**, 77, 1–34. [CrossRef]
16. Liu, B.; Zheng, Y.F.; Ruan, L. In vitro investigation of Fe30Mn6Si shape memory alloy as potential biodegradable metallic material. *Mater. Lett.* **2011**, 65, 540–543. [CrossRef]
17. Haase, H.; Rink, L. Zinc Signaling. In *Zinc in Human Health*; Ios Press: Amsterdam, The Netherlands, 2011; Volume 76, pp. 94–117.
18. García-Mintegui, C.; Córdoba, L.C.; Buxadera-Palomero, J.; Marquina, A.; Jiménez-Piqué, E.; Ginebra, M.-P.; Cortina, J.L.; Pegueroles, M. Zn-Mg and Zn-Cu alloys for stenting applications: From nanoscale mechanical characterization to in vitro degradation and biocompatibility. *Bioact. Mater.* **2021**, 6, 4430–4446. [CrossRef]
19. Vojtěch, D.; Kubásek, J.; Šerák, J.; Novák, P. Mechanical and corrosion properties of newly developed biodegradable Zn-based alloys for bone fixation. *Acta Biomater.* **2011**, 7, 3515–3522. [CrossRef]
20. Yao, C.Z.; Wang, Z.C.; Tay, S.L.; Zhu, T.P.; Gao, W. Effects of Mg on microstructure and corrosion properties of Zn–Mg alloy. *J. Alloys Compd.* **2014**, 602, 101–107. [CrossRef]
21. Baciu, E.R.; Cimpoesu, R.; Vitalariu, A.; Baciu, C.; Cimpoesu, N.; Sodor, A.; Zegan, G.; Murariu, A. Surface analysis of 3D (SLM) Co-Cr-W dental metallic materials. *Appl. Sci.* **2021**, 11, 255. [CrossRef]
22. Cimpoesu, R.; Vizureanu, P.; Stirbu, I.; Sodor, A.; Zegan, G.; Prelipceanu, M.; Cimpoesu, N.; Ioanid, N. Corrosion-Resistance Analysis of HA Layer Deposited through Electrophoresis on Ti4Al4Zr Metallic Substrate. *Appl. Sci.-Basel* **2021**, 11, 4198. [CrossRef]
23. Bejinariu, C.; Burduhos-Nergis, D.P.; Cimpoesu, N. Immersion Behavior of Carbon Steel, Phosphate Carbon Steel and Phosphate and Painted Carbon Steel in Saltwater. *Materials* **2021**, 14, 188. [CrossRef] [PubMed]
24. Cimpoesu, N.; Sandulache, F.; Istrate, B.; Cimpoesu, R.; Zegan, G. Electrochemical Behavior of Biodegradable FeMnSi-MgCa Alloy. *Metals* **2018**, 8, 541. [CrossRef]
25. Panaghie, C.; Cimpoesu, R.; Istrate, B.; Cimpoesu, N.; Bernevig, M.-A.; Zegan, G.; Roman, A.-M.; Chelariu, R.; Sodor, A. New Zn3Mg-xY Alloys: Characteristics, Microstructural Evolution and Corrosion Behavior. *Materials* **2021**, 14, 2505. [CrossRef] [PubMed]
26. Ma, S.-Y.; Liu, L.M.; Wang, S.Q. The microstructure, stability, and elastic properties of 14H long-period stacking-ordered phase in Mg–Zn–Y alloys: A first-principles study. *J. Mater. Sci.* **2014**, 49, 737–748. [CrossRef]
27. Li, L.; Jiao, H.; Liu, C.; Yang, L.; Suo, Y.; Zhang, R.; Liu, T.; Cui, J. Microstructures, mechanical properties and in vitro corrosion behavior of biodegradable Zn alloys microalloyed with Al, Mn, Cu, Ag and Li elements. *J. Mater. Sci. Technol.* **2021**, 103, 244–260. [CrossRef]
28. Yang, N.; Balasubramani, N.; Venezuela, J.; Almathami, S.; Wen, C.; Dargusch, M. The influence of Ca and Cu additions on the microstructure, mechanical and degradation properties of Zn–Ca–Cu alloys for absorbable wound closure device applications. *Bioact. Mater.* **2021**, 6, 1436–1451. [CrossRef]
29. Liu, X.; Sun, J.; Yang, Y.; Zhou, F.; Pu, Z.; Li, L.; Zheng, Y. Microstructure, mechanical properties, in vitro degradation behavior and hemocompatibility of novel Zn–Mg–Sr alloys as biodegradable metals. *Mater. Lett.* **2016**, 162, 242–245. [CrossRef]
30. Pachla, W.; Przybysz, S.; Jarzebska, A.; Bieda, M.; Sztwiertnia, K.; Kulczyk, M.; Skiba, J. Structural and mechanical aspects of hypoeutectic Zn–Mg binary alloys for biodegradable vascular stent applications. *Bioact. Mater.* **2021**, 6, 26–44. [CrossRef]
31. Watroba, M.; Mech, K.; Bednarezyk, W.; Kawalko, J.; Marciszko-Wiackowska, M.; Marzec, M.; Shepherd, D.E.T.; Bala, P. Long-term in vitro corrosion behavior of Zn-3Ag and Zn-3Ag-0.5Mg alloys considered for biodegradable implant applications. *Mater. Des.* **2021**, 213, 110289. [CrossRef]
32. Pinc, J.; Skolakova, A.; Vertat, P.; Duchon, J.; Kubasek, J.; Lejcek, P.; Vojtech, D.; Capek, J. Microstructure evolution and mechanical performance of ternary Zn-0.8Mg-0.2Sr (wt. %) alloy processed by equal-channel angular pressing. *Mater. Sci. Eng. A* **2021**, 824, 141809. [CrossRef]
33. Huang, T.; Liu, Z.; Wu, D.; Yu, H. Microstructure, mechanical properties, and biodegradation response of the grain-refined Zn alloys for potential medical materials. *J. Mater. Res. Technol.* **2021**, 15, 226–240. [CrossRef]

34. Vaclavek, L.; Tomastik, J.; Chmelickova, H.; Ctvrtlik, R. Benefits of use of acoustic emission in scratch testing. *Acta Polytech.* **2020**, *27*, 121–125. [CrossRef]
35. Gallego, A.; Piotrkowski, R.; Ruzzante, J.; Cabo, A.; Garcia-Hernandez, M.T.; Castro, E. Acoustic Emission Technique to Assess Microfractures of Metallic Coatings with Scratch-Tests. PACS Reference: 43.35.Zc. 8. Available online: http://sea-acustica.es/fileadmin/publicaciones/Sevilla02_ult03012.pdf (accessed on 24 October 2021).
36. Dambatta, M.S.; Izman, S.; Kurniawan, D.; Farahany, S.; Yahaya, B.; Hermawan, H. Influence of thermal treatment on microstructure, mechanical and degradation properties of Zn–3Mg alloy as potential biodegradable implant material. *Mater. Des.* **2015**, *85*, 431–437. [CrossRef]
37. Song, J. Theory of Diffusion Impedance in Nanostructured Electrochemical Systems. Ph.D. Thesis, Massachusetts Institute of Technology, Cambridge, MA, USA, 2019.
38. Lai, W.; Ciucci, F. Mathematical modeling of porous battery electrodes—Revisit of Newman's model. *Electrochim. Acta* **2011**, *56*, 4369. [CrossRef]
39. Roman, A.M.; Geantă, V.; Cimpoesu, R.; Munteanu, C.; Lohan, N.M.; Zegan, G.; Cernei, E.R.; Ioniță, I.; Cimpoesu, N.; Ioanid, N. In-Vitro Analysis of FeMn-Si Smart Biodegradable Alloy. *Materials* **2022**, *15*, 568. [CrossRef]

Article

Microstructural, Electrochemical and In Vitro Analysis of Mg-0.5Ca-xGd Biodegradable Alloys

Bogdan Istrate [1], Corneliu Munteanu [1,2,*], Ramona Cimpoesu [3,*], Nicanor Cimpoesu [3], Oana Diana Popescu [1] and Maria Daniela Vlad [4,*]

1. Mechanical Engineering Department, Gheorghe Asachi University of Iasi, 6 D. Mangeron Blvd, 700050 Iasi, Romania; bogdan_istrate1@yahoo.com (B.I.); diana3popescu@gmail.com (O.D.P.)
2. Technical Sciences Academy of Romania, 26 Dacia Blvd, 030167 Bucharest, Romania
3. Faculty of Material Science and Engineering Department, Gheorghe Asachi University of Iasi, 41 Dimitrie Mangeron Str., 700050 Iasi, Romania; nicanor.cimpoesu@tuiasi.ro
4. Faculty of Medical Bioengineering, Grigore T. Popa University of Medicine and Pharmacy from Iasi, 9-13 Kogălniceanu Str., 700454 Iasi, Romania
* Correspondence: cornelmun@gmail.com (C.M.); ramona.cimpoesu@tuiasi.ro (R.C.); maria.vlad@umfiasi.ro (M.D.V.); Tel.: +40-744-647-991 (C.M.); +40-743-646-660 (R.C.); +40-746-100-851 (M.D.V.)

Citation: Istrate, B.; Munteanu, C.; Cimpoesu, R.; Cimpoesu, N.; Popescu, O.D.; Vlad, M.D. Microstructural, Electrochemical and In Vitro Analysis of Mg-0.5Ca-xGd Biodegradable Alloys. *Appl. Sci.* **2021**, *11*, 981. https://doi.org/10.3390/app11030981

Received: 22 December 2020
Accepted: 15 January 2021
Published: 22 January 2021

Publisher's Note: MDPI stays neutral with regard to jurisdictional claims in published maps and institutional affiliations.

Copyright: © 2021 by the authors. Licensee MDPI, Basel, Switzerland. This article is an open access article distributed under the terms and conditions of the Creative Commons Attribution (CC BY) license (https://creativecommons.org/licenses/by/4.0/).

Abstract: The subject of Mg-based biodegradable materials, used for medical applications, has been extensively studied throughout the years. It is a known fact that alloying Mg with biocompatible and non-toxic elements improves the biodegradability of the alloys that are being used in the field of surgical applications. The aim of this research is to investigate the aspects concerning the microstructure, electrochemical response (corrosion resistance) and in vitro cytocompatibility of a new experimental Mg-based biodegradable alloy—Mg–0.5%Ca with controlled addition of Gd as follows: 0.5, 1.0, 1.5, 2.0 and 3.0 wt.%—in order to establish improved biocompatibility with the human hard and soft tissues at a stable biodegradable rate. For this purpose, scanning electron microscopy (SEM), energy-dispersive spectroscopy (EDS), light microscopy (LM) and X-ray diffraction (XRD) were used for determining the microstructure and chemical composition of the studied alloy and the linear polarization resistance (LPR) method was used to calculate the corrosion rate for the biodegradability rate assessment. The cellular response was evaluated using the 3-(4,5-dimethyltiazol-2-yl)-2,5-diphenyl tetrazolium bromide (MTT) test followed by fluorescence microscopy observation. The research led to the discovery of a dendritic α-Mg solid solution, as well as a lamellar Mg_2Ca and a Mg_5Gd intermetallic compound. The in vivo tests revealed 73–80% viability of the cells registered at 3 days and between 77 and 100% for 5 days, a fact that leads us to believe that the experimental studied alloys do not have a cytotoxic character and are suitable for medical applications.

Keywords: Mg–Ca–Gd alloys; SEM; EDS; microstructure; electrochemical evaluation; in vitro test

1. Introduction

Magnesium and its alloys are widespread in various fields such as human and veterinary medicine, and automotive and aerospace engineering, due to their superior properties in terms of biocompatibility, low modulus of elasticity, vibration damping and biodegradation capacity without side effects [1,2]. Medical applications that use biodegradable alloys based on Mg, Zn or Fe are widespread, especially for orthopedic and cardiovascular surgical applications, leading scientific research to confirm that these alloys are hierarchical in scale as opposed to their alloy successors such as stainless steel, Co–Cr alloys and Ti-based alloys [3]. Mg-based biodegradable materials are used in the manufacture of screws, plates, scaffolds, etc. due to their similar modulus of elasticity to the biological bone and the biodegradable character which contributes to the general healing process [4,5].

In addition to orthopedic applications in the form of pins, screws, rods and plates, Mg alloys have been investigated in cardiovascular applications as biodegradable Mg alloy stents [6–11] due to their biodegradability and mechanical properties. In addition, the ideal cardiovascular stent biomaterial should promote endothelialization with minimal neo-intimal hyperplasia. In this sense, any kind of injury caused by the implantation of the biomaterial into the living tissue will initiate an inflammatory reaction to the material, with fibroblasts being a key contributor to new tissue formation post-injury (i.e., depositing new collagen and facilitating healing or fibrous encapsulation), even though it may not necessarily be in direct contact with the implanted cardiovascular devices. Unfortunately, biodegradable magnesium-based alloys also have a number of disadvantages, such as low corrosion resistance and high hydrogen release during bone healing, which can cause subcutaneous inflammatory reactions [12,13]. Furthermore, the presence of various impurities that accompany the manufacturing process of these materials, such as Fe, Ni, Cu, etc., at very low concentrations leads to a decrease in biocompatibility and a lack of corrosion properties [14,15]. In order to improve these aspects, scientific studies have continued to combine Mg with biocompatible elements that do not produce toxic effects (Ca, Sr, Mn, etc.), but also rare earth (RE) elements [16]. The conceptual alloy design for these alloys with elements of the rare earth category was based on the necessity and improvement of the properties of Mg–Ca, Mg–Sr and Mg–Mn systems, etc., by increasing mechanical resistance, decreasing density and improving the rate of degradation in environments such as simulated biological fluids. Wu et al. [17] and Zhang et al. [18] established that rare earth elements are divided into two classes: light RE elements (Y, Ce, La, etc.) and heavy RE elements (Gd, Sm, Sc, etc.) and research conducted by Liu et al. identified that some of the most effective alloys used in the medical field are the Mg–Gd and Mg–Y binary alloys [19]. Gadolinium (Gd) presents a solubility of 23.49 wt.% at the eutectic temperature [20] in Mg and plays a key role in strengthening the solid solution when alloyed to Mg. Feyerabend et al. identified that Gd is one of the rare earth elements with the best biocompatibility of all REEs (rare earth elements) [21]. Hort et al. [22] noted that the Mg–Gd binary system has mechanical properties very close to those of the biological bone compared with the conventional biocompatible Co–Cr and Ti alloys. Gao et al. [23] investigated the effects of Gd on a solid solution and reported that it was an effective solid solution enhancer in Mg compared with Al and Zn. Studies conducted by Peng et al. [24] suggested that the molten Mg–20Gd alloy contained mostly the supersaturated α-Mg solid solution, while the as-cast Mg–20Gd alloy consisted of α-Mg + Mg5Gd. Nodooshan et al. [25] showed that increasing Gd will lead to the age-hardening response and tensile strength of Mg–xGd–Y–Zr (x = 3 wt.%, 6 wt.%, 10 wt.%, 12 wt.%) alloys and developed a Mg–10Gd3Y–0.5Zr alloy with an ultimate tensile strength of 390 MPa and a yield strength of 245 MPa. Gao et al. [23] published that the properties of Mg alloys had been improved by the addition of REEs due to the formation of metastable REEs containing phases along the grain boundaries. Alloying with Gd more than 10 wt.% increases the mechanical strength due to the noble behavior of Mg5Gd precipitated in grain boundaries. Research findings have also shown that Gd improves the rate of degradation of Mg alloys by stabilizing the resulting corrosion products [26], but at concentrations greater than 5 wt.%, the rate of degradation and biocompatibility could decrease. At the same time, Myrissa et al. [27] highlighted the corrosion resistance of the Mg–10Gd alloy implanted in rats and observed that alloys with a high percentage of Gd (10 wt.%) implanted for up to 36 weeks led to increased concentrations of Gd in various rat organs, such as the spleen, liver, lung and kidney. Among several other alloy elements, calcium (Ca) improves the biocompatibility and mechanical properties of magnesium-based alloys [28,29]. Additionally, the most major issue for the application of MgCa as implantable elements is still the high corrosion and electro-corrosion rate in body fluids [30–32]. Percentages higher than 1 wt.% Ca have also been shown to reduce certain characteristics such as castability, corrosion behavior and mechanical properties due to the higher volume fraction of the Mg_2Ca phases [33,34], which also has an effect on the biocompatibility of the alloy [35]. Maradze et al. presented in their studies that the report through Mg^{2+} and Ca^{2+} ions has an important part to play in the maintenance of the cellular activities of the embryonic mesenchymal cells that give rise to striated muscle fibers [36]. Upon first contact of multi-nucleated fibers with high Mg and Mg–Ca corrosion compounds, the effect was to reduce cell viability, while at the same time the myotubes became capable of adaptation. A significant percentage of Mg^{2+} ions can substantially increase cell proliferation and, at the same time, the presence of Ca^{2+} ions is extremely important for the formation of fibers [37]. The properties of Mg–Ca alloys are affected by mechanical or thermo-mechanical processing [38–40]. Therefore, alloying the Mg–Ca alloys (<1 wt.% Ca) with a third element, such as Gd, can be the right path to follow in order to improve the mechanical properties and biocompatibility of the alloy being studied in this paper. In this research paper, the authors studied a novel Mg–0.5 wt.% Ca experimental material which was alloyed with a controlled content of 0.5, 1.0, 1.5, 2.0 and 3.0 wt.% Gd in order to improve the biodegradability rate needed for a symmetrical balance between alloy degradation and bone healing, followed by in vitro 3-(4,5-dimethyltiazol-2-yl)-2,5-diphenyl tetrazolium bromide (MTT) tests to determine its biocompatibility.

2. Experimental Details

2.1. Obtaining the Magnesium with Calcium and Gadolinium Alloys' Structure and Chemical Composition Investigations

The manufacturing process of the experimental alloys used master alloys procured from Hunan China Co. [41,42] with the mass percentages shown in Table 1. The Mg-based alloys were obtained through a classical melting process using an induction furnace in an Ar-protected environment in a 675–700 °C temperature field for 30 minutes from a square section of the initial materials, as in previous research [43]. The elaboration process resulted in cylindrical mini-ingots, which were cut in smaller round plates with different chemical compositions based on the differences between the Gd element concentrations, with the values shown in Table 2. The round samples are 20 mm in diameter with a thickness of 2 mm. Material load calculations were made on the basis of the metal loss coefficient; the quantities used are shown in Table 2 (with specimen codes) with a net mass of around 23 g per melt. For experimental tests, the specimens were mechanically polished with an Al_2O_3 suspension solution (2–5 μm) after metallographic grinding with paper disks of 200–2500 MPi granulation. The surface was cleaned with ethylalcohol for 30 min and the microstructure was highlighted by chemical etching using $Mg(CH_3COO)2Mg \cdot 4H_2O$ acetate solution. Microstructural aspects were analyzed by optical microscopy (Leica DMI 5000 equipment, Leica, Wetzlar, Germany) and scanning electron microscopy (SEM)(FEI Quanta 200 3D). Chemical composition determinations were accomplished with an energy-dispersive spectroscopy (EDS) detector (Xflash, Bruker, Germany) and a X-ray diffractionsystem (XRD) (Xpert PRO MPD 3060 equipment, Panalytical, The Netherlands; copper-X-ray tube (Kα = 1.54051°), 2θ: 20°–100°).

Table 1. Magnesium and Mg-based master alloy weight percentages [41,42].

Materials	Mg/Ca/Gd (wt.%)	Fe (wt.%)	Ni (wt.%)	Cu (wt.%)	Si (wt.%)	Al (wt.%)
Pure Mg	Mg (99 wt.%)	0.15–0.2	0.17–0.2	0.14–0.2	0.15–0.2	0.16–0.2
Mg15Ca	Ca (15.29 wt.%)	0.004	0.001	0.003	0.013	0.011
Mg30Gd	Gd (28.05 wt.%)	0.010	0.001	0.001	0.006	0.011

Table 2. Initial materials masses used for metal charges in order to obtain the experimental samples.

Specimens	Chemical Composition	Mg [Grams]	Mg–15Ca (Grams)	Mg–30Gd (Grams)
MgCa0.5Gd	Mg-0.5%Ca-0.5%Gd	21.82	0.77	0.41
MgCa1Gd	Mg-0.5%Ca-1%Gd	21.42	0.77	0.82
MgCa1.5Gd	Mg-0.5%Ca-1.5%Gd	21.00	0.77	1.23
MgCa2Gd	Mg-0.5%Ca-2%Gd	20.59	0.77	1.64
MgCa3Gd	Mg-0.5%Ca-3%Gd	19.77	0.77	2.46

2.2. Electrochemical Analysis

Based on the protocol applied in [43], the experimental magnesium-based alloys' behavior was tested in simulated body fluid (SBF) electrolyte with the ion composition as shown in Table 3. An electro-corrosion resistance VoltaLab-21 potentiostat (Radiometer, Denmark) was used to analyze linear and cyclic curve determinations in SBF. The results were acquired and processed using the Volta Master 4 program package. For experiments, an experimental cell was used to expose the sample (work electrode) to an electrolyte solution at the same time with the Pt auxiliary electrode and a calomel-saturated one. The samples were isolated with Teflon, so only a specific area was exposed to the electrolyte, namely 0.78 cm^2. The solution was permanently aerated with a magnetic stirrer in order to remove the gas bubbles formed on the metal surface due to hydrogen elimination. Electrochemical polarization test methods established the criteria for susceptibility for several forms of corrosion and anodic or cathodic protection. Cyclic voltammetry, potentio-dynamic and potentiostatic techniques were necessary for evaluation of the passive region, aptitude to passivation, stability and quality of passivation (corrosion rate).

Table 3. Ions composition of the simulated body fluid (SBF) electrolyte used for experiments [44].

Electrolyte Ions (mmol/dm^3)	Na$^+$	K$^+$	Mg^{2+}	Ca^{2+}	Cl$^-$	HCO$_3^-$	HPO$_4^{2-}$	SO$_4^{2-}$
Simulated body fluid	142	5	1.5	2.5	147.8	4.2	1	0.5
Human blood plasma	142	5	1.5	2.5	103	27	1	0.5

Direct current (DC) electrochemistry, particularly the potentio-dynamic polarization test, was used to record much relevant information based on the behavior of the electrodes. Using direct current polarization mechanisms, we could obtain detailed information about the working electrodes' (MgCaGd samples) corrosion rate and the type of surface corrosion: generalized or pitting, passivation layer formation and stability, and anodic (oxidation) or cathodic (reduction) reactions. The linear polarization resistance method is probably one of the most widespread applications of electrochemical measurements in the laboratory and is widely applied in the case of uniform (generalized) corrosion.

The following coordinates were chosen by the authors for the function representation: current density [mA/cm^2] and potential [V], because this variation allows emphasis on the corrosion potential (E_0) and corrosion current (J_{corr}). The experiments were realized at room temperature (24 °C) and the potential records were registered (linear plots were registered at a scan rate of 1 mV/s and the cyclic plots at a scanning rate of 10 mV/s). The tests were repeated 4 times to achieve proper repeatability of the results. The corrosion current density, J_{corr}, can be determined using 2 equations by extrapolation of the Tafel lines: a complete polarization curve consisting of a cathodic part and an anodic part. The cathodic portion of the polarization curve contains information concerning the kinetics of the reduction reactions occurring for a particular system.

The corrosion potential is deduced from the intersection of the anodic and cathodic Tafel slopes.

This method is based on polarization resistance evaluation, Rp, which is defined as the tangent slope of the potential–current density curve [$E = f(j)$] at the equilibrium point ($E = E_0$), indicating the free corrosion potential:

$$R_p = \left[\frac{\Delta E}{\Delta j}\right]_{E=E_0} \text{ohm·cm}^2 \quad (1)$$

Corrosion current values were used to obtain the instantaneous corrosion rate: V_{corr} (µm/year):

$$V_{corr} = 22.9 \cdot J_{corr}, \quad (2)$$

$$J_{corr} = \frac{b_a \cdot b_c}{2.303(b_a + b_c) \cdot R'_p} (\text{mA/Scm}^2) \quad (3)$$

2.3. Cytocompatibility Analysis

2.3.1. Alloy Sample Preparation

The alloy specimens with the appearance of flat samples (with a mean weight of 0.83 g) were sonicated in an acetone–ethanol mix in order to eliminate impurities, sterilized under ultraviolet (UV) light for a time interval of 30 min on each side (3 samples for each studied alloy) and subsequently placed in well-permeable supports (with a 0.4-µm pore membrane insert) in order to co-incubate them with the cells for the cellular viability study.

2.3.2. Cell Culture

For the cell viability study, we used albino rabbit dermal primary fibroblasts (passage No. 3) cultured in a complete Dulbecco's modified Eagle medium/Nutrient F-12 Ham (DMEM-F12 Ham) culture medium. For this, 10% inactivated fetal bovine serum (FBS) and 1% antibiotic solution (containing 5000 units penicillin, 5 mg streptomycin and 10 mg neomycin/mL) were added to the cell culture media prior to use, and the cells were kept under standard cell culture conditions (i.e., a humid incubator at 37 °C and 5% CO_2). Briefly, after attaining the 90% confluence level, the cells were washed with phosphate-buffered saline (PBS) solution, detached by incubation with trypsin/EDTA (Sigma Chemical Co., St. Louis, MA, USA), used for seeding in 24-well culture plates at a density of 1×10^4 cells/well and then incubated for 24 h to facilitate cell adhesion on the wells' bottom. Subsequently, the medium was changed with complete fresh medium, after eliminating the unattached or dead cells by washing with PBS [44,45]. Afterwards, the experimental Mg alloy samples placed in well-permeable supports were suspended on the top of the 24-well-plated cell cultures and co-incubated under the abovementioned conditions for 1, 3 and 5 days. The medium was refreshed every 2 days. A similar methodology was applied for both the cytocompatibility assay and the cell morphology evaluation [46]. It should be pointed out that our method favored the interaction/reaction between the studied Mg alloys with the cell culture medium [47]. In this study, we used control wells (i.e., negative controls) containing cell cultures only (i.e., 3 control wells for each time period).

2.3.3. Cell Viability

The alloys' cytocompatibility was evaluated by the 3-(4,5-dimethyltiazol-2-yl)-2,5-diphenyl tetrazolium bromide (MTT) colorimetric assay [35], similar to in the literature [37,38], which allows quantification of the live cells' metabolic activity. The cell viability study was done through direct co-incubation of the cells with the studied Mg alloy samples (i.e., continuous cell exposition to an 100% extract) but avoiding cell mechanical injury as a consequence of the Mg alloys disintegrating by dangling the permeable supports, containing the metal samples, into the 24-wellplates. Briefly, after established incubation periods (i.e., 1, 3 and 5 days), the permeable supports were removed from the wells and the cells were treated with a MTT dye solution (for 3 h at 37 °C) in order to facilitate intracellular formazan crystal formation. Subsequently, the solubilization of the insoluble dark-blue formazan product was performed with isopropyl alcohol by continuous agitation for 15 min using an Environmental Shaker-Incubator (type ES-20, Biosan, Riga, Latvia). The characteristic absorbance of each well was read at a wavelength of 570 nm using a microplate reader (Tecan Sunrise, with Magellan V.7.1 data processing software). The cell viability data for the studied Mg alloys were expressed as a percentage of the cell viability data related to the controlwells by means of the formula: cellular viability = 100 × (alloy wells' optical density—empty wells' optical density)/(control wells' optical density—empty wells' optical density). The one-way ANOVA test was used for statistical analysis of cell viability data; Tukey's method was used for comparing the obtained results and statistically significant differences were accepted at $p < 0.05$.

2.3.4. Cell Morphology

For fluorescence staining, after co-incubation periods of 1 and 5 days (under the conditions explained in the sections above), the cells in each well were rinsed with Hanks' Balanced Salt solution (HBSS) (H8264, Sigma-Aldrich, St. Louis, MO, USA) without red phenol, then 200 µL of a 1:1000 calcein solution (Calcein AM; C1359, Sigma-Aldrich)was added in HBSS. After incubation for 30 min, in the dark and at 37 °C, the cells' morphology and distribution were investigated through visualization and image acquisition using an inverted microscope (Leica DMIL LED equipped with a Leica DFC450C camera and Leica Application Suite—Version 7.4.1 image acquisition software, Leica, Wetzlar, Germany).

3. Results and Discussions

3.1. Microstructural Analysis

The morphological and microstructural analysis of Mg–0.5Ca–xGd alloys by light microscopy is presented in Figure 1. The microstructure typically exhibits aspects of as-melted metallic alloys with Mg_2Ca tabular intermetallic structures and particles of Mg_5Gd which are proportionally equably distributed. The enhancement with Gd shows the growth of round particles with a separation trend of somewhat uniformly distributed black dots [48]. The structural appearance of the SEM images of the said specimens is shown in Figure 2 and the existence of an intermetallic stage is noticeable. Gd-based compounds are usually acicular and round in shape and are observed to be whiter than the Mg2Ca tabular intermetallic compounds. Gd has a slightly low influence on the size and shape of Mg grains, confirmed by literature studies [49,50] on Mg-based alloys with a Gd content between 1 wt.% and 4 wt.%.

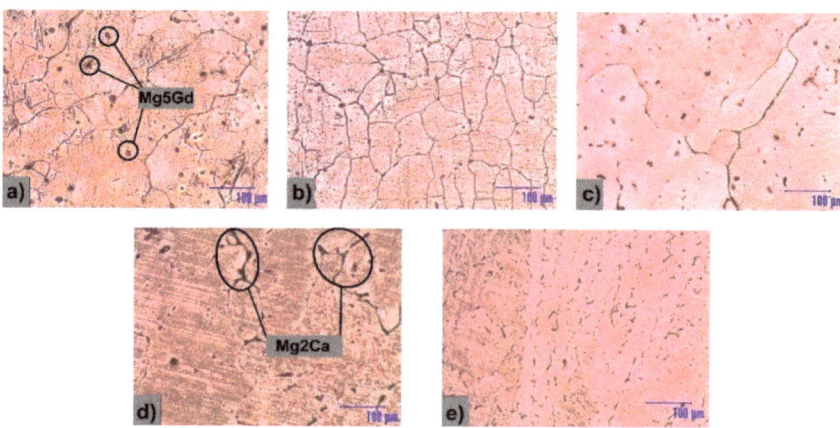

Figure 1. Optical images of samples: (**a**) Mg–0.5Ca–0.5Gd; (**b**) Mg–0.5Ca–1Gd; (**c**) Mg–0.5Ca–1.5Gd; (**d**) Mg–0.5Ca–2Gd; (**e**) Mg–0.5Ca–3Gd.

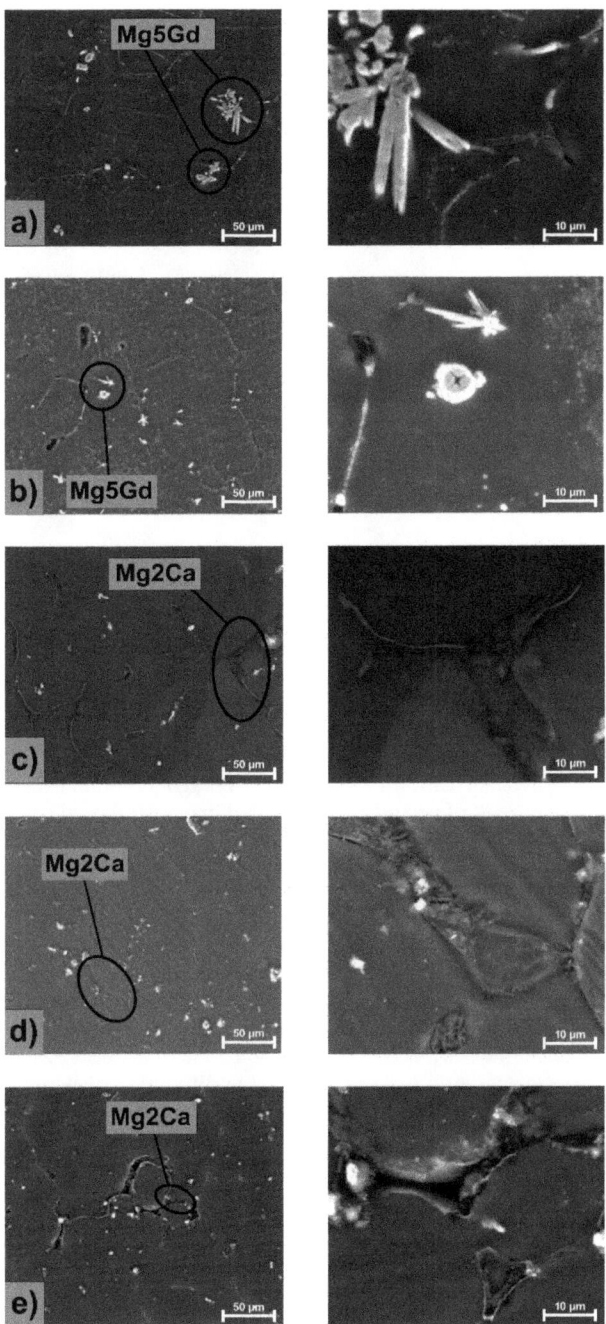

Figure 2. SEM images of the samples: (**a**) Mg–0.5Ca–0.5Gd; (**b**) Mg–0.5Ca–1Gd; (**c**) Mg–0.5Ca–1.5Gd; (**d**) Mg–0.5Ca–2Gd; (**e**) Mg–0.5Ca–3Gd.

The chemical composition of the Mg–0.5Ca–xGd alloys was investigated by energy dispersive spectroscopy and is shown in Figure 3. Cross-sectional areas of the specimens were analyzed in 10 separate zones and the mean values are given.

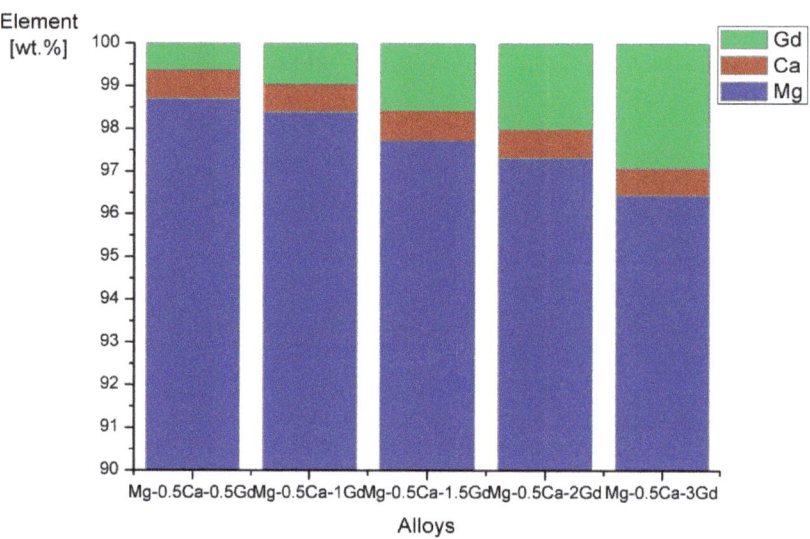

Figure 3. Average elemental composition obtained by energy-dispersive spectroscopy (EDS) analysis (SD Mg: ±0.27, ±0.45, ±0.26, ±0.36, ±0.38; SD Ca: ±0.11, ±0.08, ±0.11, ±0.12, ±0.12; SD Gd: ±0.13, ±0.18, ±0.16, ±0.26, ±0.41).

The structural morphology investigation emphasizes homogeneous elements and the growth of peculiar phases, rated in a α-Mg-based solid solution (α-Mg), Mg_2Ca and Mg_5Gd. The Mg_2Ca complex element is situated at the limits of the α-Mg grains, realizing a film-coated eutectic with α-Mg. X-ray diffraction peaks of the experimentally Mg–0.5Ca–xGd alloys are given in Figure 4. From the XRD analysis shown in Figure 4, the Mg, Mg_2Ca and Mg_5Gd phases were detected in the X-ray diffraction peaks of the experimentally Mg–0.5Ca–xGd alloys. α-Mg (ICDD-PDF-International Center for Diffraction Data-PDF: 01-071-9399) has been identified at different 2θ values (30.24°, 35.77°, 46.82°, 56.34° and 77.62°) as the main structure with a hexagonal crystalline phase. Moreover, the existence of Mg_2Ca (ICDD-PDF: 01-073-5122) was shown at 2θ = 34.86°, 67.42° and 96.88°, and Mg_5Gd (ICDD-PDF: 01-071-9618) in a cubic system at 2θ = 29.46°, 61.15° and 87.64°. As the Gd percentage gradually increased, the diffraction peak of Mg_5Gd intensified. The lattice parameters of the Mg–0.5Ca–xGd alloy phases are presented in Table 4.

Table 4. Lattice parameters of Mg–0.5Ca–xGd alloy phases.

Compound	Space Group	Crystal System	a (Å)	b (Å)	c (Å)	α (°)	β (°)	γ (°)	Cell Volume (10^6 μm³)	RIR
Mg	P63/mmc	Hexagonal	3.2089	3.2089	5.2101	90	90	120	46.46	4.01
Mg_2Ca	P63/mmc	Hexagonal	6.2250	6.2250	10.1800	90	90	120	341.63	2.30
Mg_5Gd	F43m	Cubic	3.9500	3.9500	3.9500	90	90	90	61.63	18.05

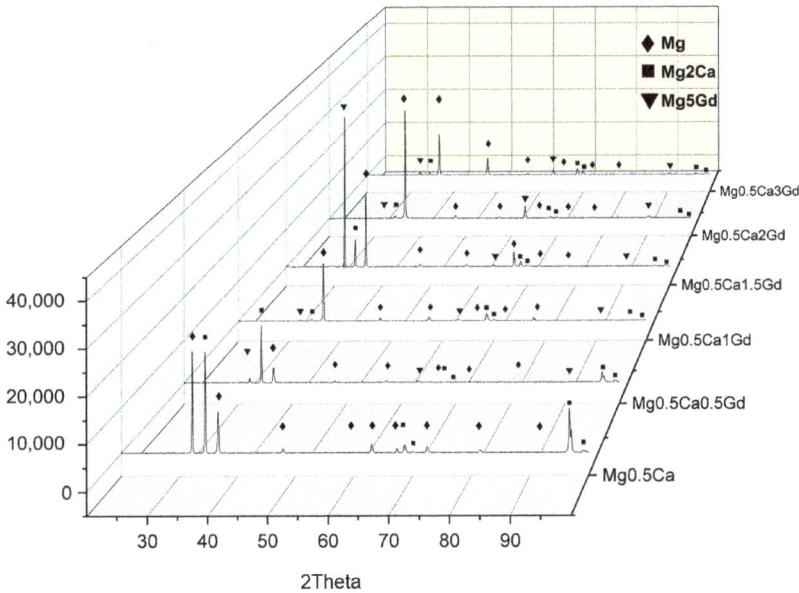

Figure 4. XRD pattern of Mg–0.5Ca–xGd samples.

3.2. Electrochemical Corrosion Resistance Analysis

Normally, in the process of electrochemical corrosion, the cathodic polarization curve is assumed to represent the cathodic H_2 evolution through solution reduction, while the anodic polarization curve represents the dissolution of the magnesium alloy.

The alloys in contact with the electrolyte solution present both anodic and cathodic reactions, with higher activity for oxidation process (in all cases $b_a > b_c$), with an increase in the difference at the beginning of Gd alloying (first three samples) and a drop in the differences between anodic and cathodic reactions for samples with 2 or 3 wt.% Gd. The highest oxidation rate was observed in the Mg–0.5Ca–1.5Gd samples and the lowest with the reduction reactions for the Mg–0.5Ca–0.5Gd sample. However, the rate of corrosion depended on the intensity of both reactions [51].

The experimental conditions of the electro-corrosion experiment realized in SBF electrolyte are given in Table 5. Because magnesium and magnesium-based alloys discharge large amounts of hydrogen into the exposed solution part of the sample, the gaseous gaps appear to be in the form of continuous bubbles [52]. Removal of the H_2 voids from the metallic surface was realized with a magnetic stirrer at a controlled speed; in our case, we used a much-reduced rate of SBF shuffling in order to avoid the acceleration of the corrosion rate. After every determination, based on the fact that the electrolyte solution changed its color, a new quantity of SBF was used. At the contact with the SBF electrolyte solution, the metallic specimen presented an oxidation process to metal ions based on the following reactions:

$$Mg \rightarrow Mg^{2+} + 2e^- \qquad (4)$$

$$Ca \rightarrow Ca^{2+} + 2e^- \qquad (5)$$

The electrons from the anodic reaction were consumed by a corresponding cathodic Reaction (4) and the reduction of oxygen dissolved in H_2O.

$$2H_2O + 2e^- \rightarrow H_2 + 2OH^- \qquad (6)$$

Table 5. Parameters obtained from the electro-corrosion resistance tests of the experimental alloys Mg–0.5Ca–xGd (x = 0.5; 1; 1.5; 2 and 3 wt.%).

Sample	E_0 mV	b_a mV	b_c mV	R_p ohm·cm^2	J_{cor} µA/cm^2	V_{cor} mm/Y	OCP V
Mg–0.5Ca	−3511.2	217.8	−188.6	0.114	170.11	4.2	−1.72
Mg–Ca–0.5Gd	−1740.8	283.8	−240.4	0.993	42.42	1.0	−1.56
Mg–Ca–1 Gd	−2002.8	307.4	−265.0	1.54	32.2	0.8	−1.62
Mg–Ca–1.5 Gd	−1576.0	264.1	−250.2	2.55	16.53	0.41	−1.68
Mg–Ca–2 Gd	−1491.9	196.9	−179.8	2.76	12.71	0.28	−1.61
Mg–Ca–3 Gd	−1447.7	256.1	−236.3	2.25	5.92	0.16	−1.58

The compounds formed on the surface after immersion was based on the alloys' main elements, especially magnesium and calcium oxides. At the beginning of the contact between the experimental alloys and the electrolytes, a protective oxide layer appeared on the surface and, for a certain period of time, provided protection against corrosion; the layer was generally formed from $Mg(OH)_2$. The formation and growth of the $Mg(OH)_2$ layer was followed by a hydrogen discharge, as directly observed by visual inspection. As a conclusion, the formed $Mg(OH)_2$ layer cannot provide effective protection to the Mg substrate, at least not for a longer period of time. The pH of the solution varied with the chemical reactions that occurred on the surface as a basic general characteristic and sometimes, especially with the formation of HCl, decreased due to the influence of acid formation.

The corrosion current (J_{corr}) expresses the reduction value of the experimental materials in contact with an electrolyte solution. The corrosion density presented different values for the experimental alloys from 5.92 µA/cm^2 to 170.11 µA/cm^2 (Table 5).

The influence of the gadolinium element added to the magnesium–calcium alloy on the electro-corrosion resistance properties of the system is highlighted in Table 5. A reduction in the corrosion rate of up to 20 times with the addition of 3 wt.% of Gd was observed compared with the Mg–0.5Ca-based experimental alloy, Mg_0.5Ca. In quantitative terms, the percentage of Gd significantly influenced the behavior of the alloy at values higher than 2 wt.% of Gd. It is known that both intermetallic compounds have a significant influence on the electro-corrosion resistance of Mg-based materials [48,49]. The open circuit potential (OCP) shows close values for all samples, decreasing with the elevated percentage of the alloying element. The open circuit potential (OCP) usually varies with the immersion time because of the changes in the nature of the sample's surface during its exposure to a corrosive medium. The lower OCP value in the case of Mg–Ca–3Gd confirms the formation of a passivation layer on the sample's surface. This shows that the intermetallic compounds retain the cathodic reactions more than anodic, and positive ions have time to diffuse. The Mg–0.5Ca alloy had a more negative potential (−3511.2 mV) than the Mg–Ca–xGd alloy (−1447.7 mV), indicating that the addition of Gd shifted to a more noble direction, thus having a positive effect on the corrosion resistance of the alloys. It can be observed that the cathodic polarization in the Mg–0.5Ca–1Gd alloy was significantly higher than that of the Mg–0.5Ca–2Gd, indicating that the corrosion potential (E_0) was more negative in the alloy with reduced amounts of Gd. It can also be seen that the anodic polarization curve of the Mg–Ca–1Gd alloy was greater than that of the Mg–Ca alloy. The lower J_{corr}, along with the longer anodized polarization curve and more negative E_0 results, demonstrates that the corrosion resistance was improved for the Mg–Ca–Gd alloy (Figure 5). The corrosion current density of the Mg–0.5Ca alloy was 170 µA/cm^2, which is significantly higher than that on Mg–Ca–3Gd (5.92 µA/cm^2). As a result, the formation of a specific phase, namely the α-Mg-based Mg$_5$Gd solid solution had a higher influence on the corrosion behavior of the alloys.

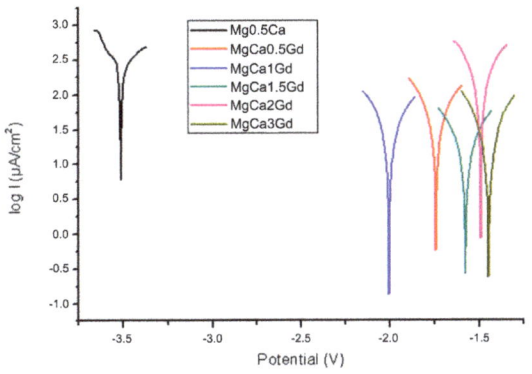 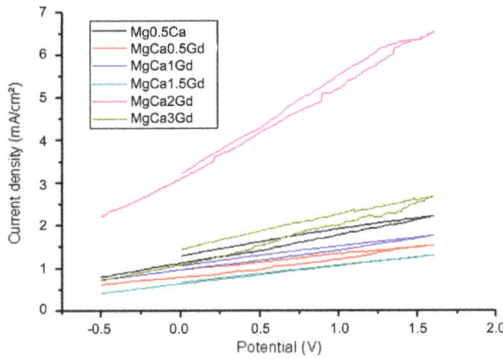

Figure 5. Tafel diagrams of Mg–0.5Ca–xY-based experimental alloys.

During cyclo-voltammetric sweeps (Figure 5) changes developed in the passive films and in the solution composition at the metal–electrolyte interface (e.g., hydroxide formation and pH increase), as well as development of the general corrosion form.

All specimens had a generally corroded surface, particularly due to linear and cyclic polarization tests. In all cases, in addition to the detachment of the material from the surface of the alloys, the growth of the different compounds took place through contact between the surface of the experimental materials and the electrolyte. Corrosion compounds appeared on the surface of magnesium-based alloys, exhibiting many micro-cracks and various morphological details.

The polarization cyclic curves present the generalized corrosion of the surface of all specimens. Cyclic voltammetry consists of scanning of the electrode potential in the positive direction up to a determined value for current or potential, after which, the scanning is immediately reversed towards more negative values until it reaches the initial potential; in some cases, this scanning is repeated to determine the modifications produced on the current–potential curve obtained by scanning. However, in our case, the cyclic voltammograms no longer show a hysteresis loop; the cathode branch of the voltammogram practically overlaps the anodic branch, except for a narrow potential range located in the immediate vicinity of the potential from which dissolution starts. This little "backwards" shift of the cathode branch may be explained by the modification of the specimens' surface following corrosion (as a consequence of corrosion, the metal surface increases, making the current density decrease). The very small area of the cyclic curve loop reflects an insignificant increase of zonal corrosion (pitting type).

Figure 6 shows SEM images of the surface of experimental Mg–0.5Ca–xGd after the electro-corrosion resistance test with different image magnification powers. A generalized corrosion surface is observed, covered with reaction compounds that passed from the electrolyte solution (SBF) to the surface.

In addition to the layer of magnesium oxide formed on the surface of the alloy, there were several different formations of compounds that could be oxides of other elements, carbonates or salts (sodium or potassium chlorides). The compounds that covered the alloy formed a layer that insulated the alloy from the electrolyte medium. Compounds formed on the surface had numerous cracks and fissures that confirm the instability of these compounds on the surface. Figure 6a shows the typical morphology of magnesium oxide. The compounds on the surface have different morphologies, which indicate the possibility of forming compounds with different chemical compositions. Figure 6b shows the identification of the chemical elements that appear on the corroded surface of the Mg–0.5Ca–0.5Gd alloy.

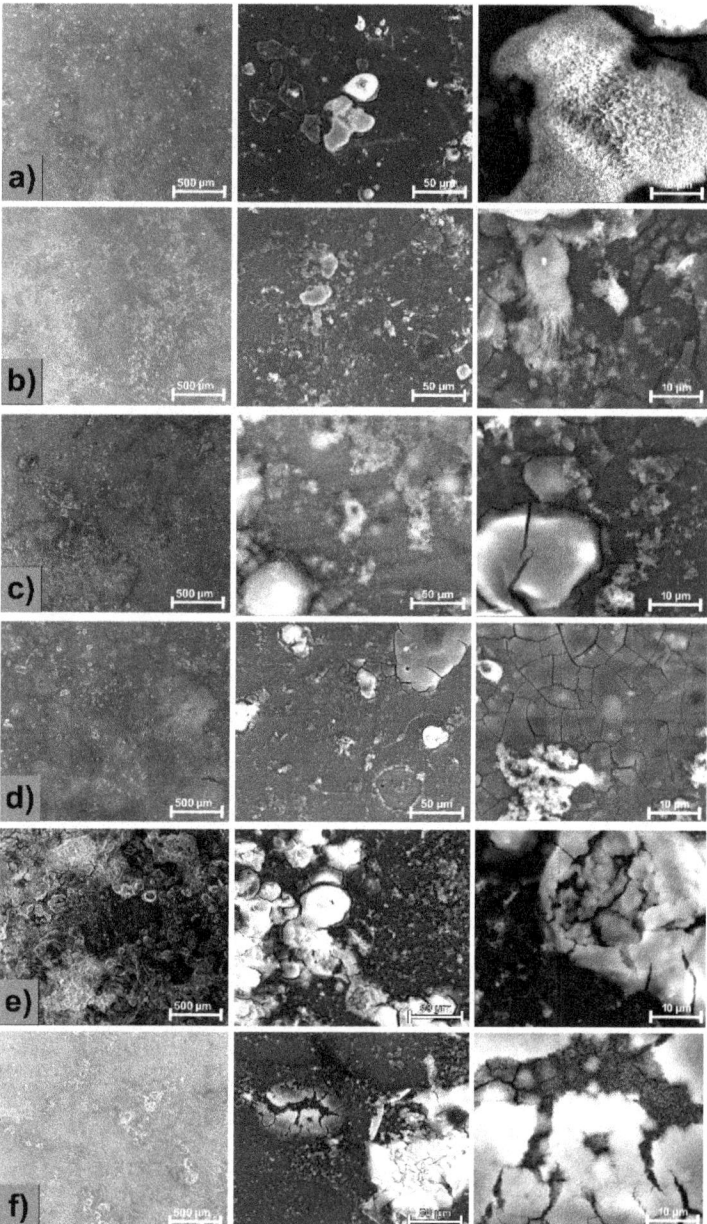

Figure 6. SEMs of the samples surfaces after electrochemical tests: (**a**) Mg–0.5Ca; (**b**) Mg–0.5Ca–0.5Gd; (**c**) Mg–0.5Ca–1Gd; (**d**) Mg–0.5Ca–1.5Gd; (**e**) Mg–0.5Ca–2Gd; (**f**) Mg–0.5Ca–3Gd.

Surface oxidation in all cases was confirmed by the chemical composition determinations (Table 6) using the EDS detector (values are average of five determinations from 1-mm^2 areas). Sample Mg–Ca–2Gd presents higher stability of the oxides on the surface and, together with samples Mg–Ca–1.5Gd and Mg–Ca–1Gd, also has traces of chloride and sodium as salts or other compounds.

Table 6. Elements identified on the Mg–0.5Ca–xGd alloys' surface after the electro-corrosion resistance tests.

		Mg wt.%	Ca wt.%	Gd wt.%	O wt.%	Cl wt.%	Na wt.%
Mg–0.5Ca	Chemical elements						
	Surface with oxides	81	0.8	-	18.2	-	-
Mg–0.5Ca–0.5Gd	Chemical elements						
	Surface with oxides	70.5	0.8	0.8	27.9	-	-
Mg–0.5Ca–1Gd	Chemical elements						
	Surface with oxides	77.7	1.1	1.2	18.7	-	1.3
Mg–0.5Ca–1.5Gd	Chemical elements						
	Surface with oxides	72.2	0.8	0.8	23.3	2.2	0.6
Mg–0.5Ca–2Gd	Chemical elements						
	Surface with oxides	65.3	0.9	0.6	32.2	0.3	0.7
Mg–0.5Ca–3Gd	Chemical elements						
	Surface with oxides	83.1	0.5	1.3	15.1		

SD (determined for the main elements after 20 chemical composition analyses on the same area) are: Mg: ±1.1; Ca: ±0.2; Gd: ±0.1; O: ±1.5; Cl: ±0.1; Na: ±0.1.

The main elements identified (with percentages higher than 0.01%) are: oxygen, magnesium, calcium and gadolinium. The elements calcium, gadolinium and chlorine appear, which implies the participation of these elements in two or more bonds characteristic of different compounds. The increase in Gd content at values greater than 2% is a success in terms of corrosion resistance, confirmed by the Tafel linear results. In the case of the 3% Gd alloy, the presence of O, Na, C and Cl elements was identified on the surface by the interaction of the experimental alloy with the SBF electrolyte solution. The presence of chlorine-based salts is also noted.

3.3. Cytocompatibility Study

In the current study, the MTT test results for the cytocompatibility assessment of Mg–0.5Ca–xGd experimental alloys are shown in Figure 7, calculated as percentages from the viability of the negative control (i.e., control wells). The results of Day 1 of co-incubation revealed that the level of cell viability varied in an increasing way and proportional to the increase in the percentage of Gd as the alloying element, being similar to the case of alloys with 2% Gd and 3% Gd ($p > 0.05$), but significantly higher compared with the viability level results in the case of the other experimental samples ($p < 0.05$). The cytocompatibility profile recorded after 3 days of co-incubation with the studied alloys reached a similar level and showed that there were no statistically significant differences among all studied samples ($p > 0.05$) at 3 days. It was observed that with an increase in co-incubation time for up to 5 days, the cell viability profile registered an increase in the case of all alloys but remained significantly lower ($p < 0.05$) for alloys with 0.5% Gd and 1% Gd. In addition, the results showed no significant differences in cell viability after 5 days for samples with a Gd quantity of 1.5%, 2% and 3% ($p > 0.05$), similar to the Mg–Ca control alloy. The MTT results seem to suggest the fact that increasing the percentage of gadolinium used for alloying did not affect the viability level in a dose-dependent manner (i.e., no significant differences among the 1.5, 2 and 3% Gd-alloyed Mg-Ca alloys) and this could be due to an increase in the stability of alloys in a humid environment (as was detailed in the previous section; i.e., Electrochemical Corrosion Resistance Analysis) with a complex composition (such as culture medium or simulated biological fluids). The decreased level of viability after 1day of co-incubation of the cells with the alloy samples could be attributed to the degradation of the alloys immersed in the experimental culture medium [53], which has the immediate consequence of releasing hydrogen with a change in pH (i.e., subsequent alkalinization, as a consequence of an Mg^{2+}-related anodic reaction by which H_2O is reduced to H_2 and OH^-). As other authors have indicated, Mg alloy degradation events may harmfully affect the cell metabolism through ion bulk leakage from Mg alloys and subsequent precipitation of salts with an inhibitory or toxic influence, and by increasing of the osmolarity (as a result of alloy reactivity) which may lead to hyperosmotic shock [54]. It should be point out that because of Mg alloys' degradation depend on the microenvironment conditions [55], different corrosion/degradation events under particular biomedical scenarios (i.e., different flow and/or composition of biological fluids) can take place by affecting the surface chemistry of the implants. These facts are still to be proved by an ongoing in vivo study.

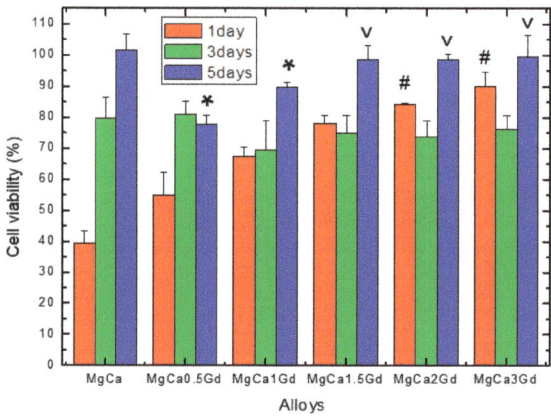

Figure 7. Cell viability vs. studied Mg–0.5Ca–xGd samples: Effect of Gd alloying on cell viability after 1, 3, and 5 days of co-incubation. Results are shown as percent of the control wells' viability (i.e., negative control). (∨, #) Not statistically significant ($p > 0.05$); (*) statistically significant ($p < 0.05$).

Furthermore, the level of viability recorded at 3 days was 73–80% and, at 5 days, was 77–100%, which permits us to consider (according to the ISO 10993-5 standard [56]) that the experimental samples do not appear to exhibit a long-term cytotoxic character. This is due to the continuous exposition of the cells to the sum of the abovementioned occurrences that occurred during the biodegradation of Mg alloys (i.e., exposure to 100% extracts for up to 5 days).

The registered viability profiles correlate with the results on the density and morphology of fibroblasts (Figure 8) co-incubated with the experimental samples for 24 h and 5 days, respectively. It has been shown that after 24 h, the cell density in the case of the experimental alloys was lower than in the case of the control wells (i.e., negative control, containing only cultured cells without the Mg alloy sample). This finding is in agreement with the cell viability results, being confirmed by the data obtained after 5 days of co-incubation of the cells with the experimental alloys. In addition, different cell spreading was observed inside the wells, such as high density areas containing cells having an elongated appearance and a bipolar morphology, and low density areas (observed towards the center of the wells and around the inserted membrane), where the cells had a polygonal shape with extensive lamellipodia-like cytoplasmic processes. The cells' variable morphology maybe associated with the regional micro-environmental changes inside the wells concerning the Ca, Mg and Gd ionic concentrations (as a result of the degradation of Mg alloys inside the insert), considering the associated structural role of calcium in the reorganization of cytoskeleton constituents [57], while magnesium accumulated up to a critical concentration could have an inhibitory effect [58].

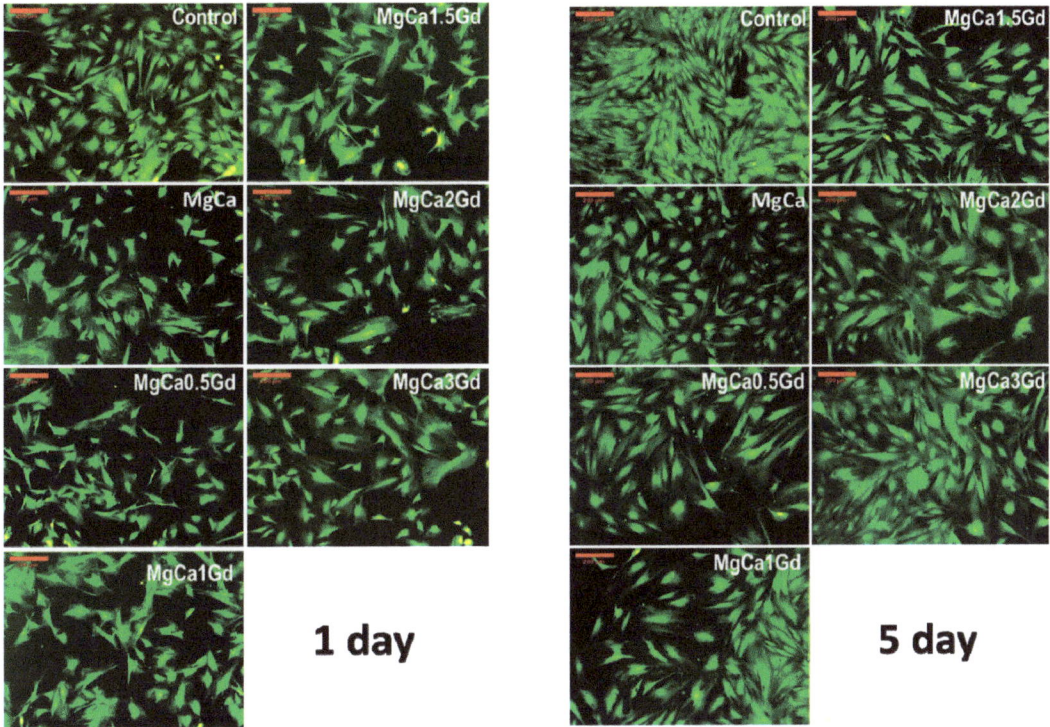

Figure 8. Fibroblastic cells' morphology evaluated by fluorescence microscopy after 1 and 5 days of co-incubation with the studied samples. Viable cells are stained in green as a result of calcein fluorescent dye's presence inside the cells. Bar: 200 μm.

4. Conclusions

In this article, five experimental biodegradable Mg-based alloys (Mg–0.5Ca–0.5Gd, Mg–0.5Ca–1Gd, Mg–0.5Ca–1.5Gd, Mg–0.5Ca–2Gd and Mg–0.5Ca–3Gd) were analyzed. In summary, the authors highlighted the influence of Gd in a Mg–0.5Ca system from the point of view of microstructure, corrosion resistance and cell viability. Microstructural characterization results from optical and scanning electron microscopy and XRD analysis highlighted patterns of a refined polyhedrical shape of Mg grains, with cubic crystalline structures, due to Gd addition. The morphological aspects of Gd particles (Mg_5Gd—white compounds) seem to have an acicular or round aspect and at the Mg grains' limit, the Mg_2Ca eutectic compound is formed due to Ca addition. The results on the degradation rate performed in a simulated body fluid solution highlighted low values and showed an area slightly affected by corrosion. The best results of electro-corrosion resistance were achieved by the Mg–0.5Ca–3Gd, which showed a degradation rate of 0.28 mm/yr. Decreasing the Gd concentration led to increasing the corrosion rate up to 4.2 mm/yr (0.5 wt.%). The presented experimental samples highlighted very good cell viability at 3 days (73–80% range) and at 5 days (77–100% range). Some low viability results were obtained for the experimental alloys (0.5 wt.%–2 wt.% Gd) after 24 h, a fact which should be attributed to the samples' reactivity leading to ion release from the alloys, pH and osmolarity increasing, and salt precipitation with toxic or inhibitory effects. Cell viability after 5 days was not significantly different for alloys with a Gd amount of 1.5%, 2% and 3% ($p > 0.05$) compared with the MgCa control alloy, so increasing the Gd alloying amount also improved the cytocompatibility of these alloys (as a consequence of improving the alloys' stability). The study will continue with in vivo analysis, in order to see the exact behavior of the experimental alloys implanted within laboratory animals.

Author Contributions: Conceptualization, B.I., C.M., N.C. and M.D.V.; methodology, C.M., R.C. and M.D.V.; software, B.I. and M.D.V.; validation, C.M., R.C., N.C. and M.D.V.; formal analysis, O.D.P.; investigation, B.I., R.C. and M.D.V.; resources, C.M. and N.C.; data curation, B.I. and M.D.V.; writing—original draft preparation, B.I., R.C. and M.D.V.; writing—review and editing, C.M., R.C. and M.D.V.; visualization, N.C.; supervision, C.M. and N.C. project administration, C.M.; funding acquisition, C.M. All authors have read and agreed to the published version of the manuscript.

Funding: This work was supported by a grant from the Romanian Ministry of Research and Innovation, CCCDI—UEFISCDI, project number PN-III-P1-1.2-PCCDI-2017-0239/60PCCDI 2018, within PNCDI III.

Institutional Review Board Statement: Not applicable.

Informed Consent Statement: Not applicable.

Data Availability Statement: Not applicable.

Conflicts of Interest: The authors declare no conflict of interest.

References

1. Ali, Y.; Qiu, D.; Jiang, B.; Pan, F.; Zhang, M.-X. Current research progress in grain refinement of cast magnesium alloys: A review article. *J. Alloys Compd.* **2015**, *619*, 639–651. [CrossRef]
2. Song, J.; She, J.; Chen, D.; Pan, F. Latest research advances on magnesium and magnesium alloys worldwide. *J. Magnes. Alloy.* **2020**, *8*, 1–41. [CrossRef]
3. Kiani, F.; Wen, C.; Li, Y. Prospects and strategies for magnesium alloys as biodegradable implants from crystalline to bulk metallic glasses and composites—A review. *Acta Biomater.* **2020**, *103*, 1–23. [CrossRef] [PubMed]
4. Höhn, S.; Virtanen, S.; Boccaccini, A.R. Protein adsorption on magnesium and its alloys: A review. *Appl. Surf. Sci.* **2019**, *464*, 212–219. [CrossRef]
5. Yazdimamaghani, M.; Razavi, M.; Vashaee, D.; Moharamzadeh, K.; Boccaccini, A.R.; Tayebi, L. Porous magnesium-based scaffolds for tissue engineering. *Mater. Sci. Eng.* **2017**, *71*, 1253–1266. [CrossRef] [PubMed]
6. Mario, C.D.; Griffiths, H.; Goktekin, O.; Peeters, N.; Verbist, J.; Bosiers, M.; Deloose, K.; Heublein, B.; Rohde, R.; Kasese, V.; et al. Drug-eluting bioabsorbable magnesium stent. *J. Interv. Cardiol.* **2004**, *17*, 391–395. [CrossRef] [PubMed]
7. Zartner, P.; Cesnjevar, R.; Singer, H.; Weyand, M. First successful implantation of a biodegradable metal stent into the left pulmonary artery of a preterm baby. *Catheter. Cardiovasc. Interv.* **2005**, *66*, 590–594. [CrossRef] [PubMed]
8. Waksman, R.; Pakala, R.; Kuchulakanti, P.K.; Baffour, R.; Hellinga, D.; Seabron, R.; Tio, F.O.; Wittchow, E.; Hartwig, S.; Harder, C.; et al. Safety and efficacy of bioabsorbable magnesium alloy stents in porcine coronary arteries. *Catheter. Cardiovasc. Interv.* **2006**, *68*, 607–617. [CrossRef]
9. Lim, G.B. Interventional cardiology: DREAMS of a bioabsorbable stent coming true. *Nat. Rev. Cardiol.* **2013**, *10*, 120. [CrossRef]
10. Heublein, B.; Rohde, R.; Kaese, V.; Niemeyer, M.; Hartung, W.; Haverich, A. Biocorrosion of magnesium alloys: A new principle in cardiovascular implant technology. *Heart* **2003**, *89*, 651–656. [CrossRef]
11. Panahi, Z.; Tamjid, E.; Rezaei, M. Surface modification of biodegradable AZ91 magnesium alloy by electrospun polymer nanocomposite: Evaluation of in vitro degradation and cytocompatibility. *Surf. Coat. Technol.* **2020**, *386*, 125461. [CrossRef]
12. Toong, D.W.Y.; Ng, J.C.K.; Huang, Y.; Wong, P.E.H.; Leo, H.L.; Venkatraman, S.S.; Ang, H.Y. Bioresorbable metals in cardiovascular stents: Material insights and progress. *Materialia* **2020**, *12*, 100727. [CrossRef]
13. Witte, F.; Hort, N.; Vogt, C.; Cohen, S.; Kainer, K.U.; Willumeit, R.; Feyerabend, F. Degradable biomaterials based on magnesium corrosion. *Curr. Opin. Solid State Mater. Sci.* **2008**, *12*, 63–72. [CrossRef]
14. Waizy, H.; Seitz, J.M.; Reifenrath, J.; Weizbauer, A.; Bach, F.-W.; Lindenberg, A.M.; Denkena, B.; Henning, W. Biodegradable magnesium implants for orthopedic applications. *J. Mater. Sci.* **2013**, *48*, 39–50. [CrossRef]
15. Song, G.L.; Atrens, A. Corrosion Mechanisms of Magnesium Alloys. *Adv. Eng. Mater.* **1999**, *1*, 11–33. [CrossRef]
16. Zheng, Y.F.; Gu, X.N.; Witte, F. Biodegradable metals. *Mater. Sci. Eng. R. Rep.* **2014**, *77*, 1–34. [CrossRef]
17. Wu, G.; Wang, C.; Sun, M.; Ding, W. Recent developments and applications on high-performance cast magnesium rare-earth alloys. *J. Magnes. Alloy.* **2020**. [CrossRef]
18. Zhang, J.; Liu, S.; Wu, R.; Hou, L.; Zhang, M. Recent developments in high-strength Mg-RE-based alloys: Focusing on Mg-Gd and Mg-Y systems. *J. Magnes. Alloy.* **2018**, *6*, 277–291. [CrossRef]
19. Liu, J.; Bian, D.; Zheng, Y.; Chu, X.; Lin, Y.; Wang, M.; Lin, Z.; Li, M.; Zhang, Y.; Guan, S. Comparative in vitro study on binary Mg-RE (Sc, Y, La, Ce, Pr, Nd, Sm, Eu, Gd, Tb, Dy, Ho, Er, Tm, Yb and Lu) alloy systems. *Acta Biomater.* **2020**, *102*, 508–528. [CrossRef]

20. Tekumalla, S.; Seetharaman, S.; Almajid, A.; Gupta, M. Mechanical Properties of Magnesium-Rare Earth Alloy Systems: A Review. *Metals* **2015**, *5*, 1–39. [CrossRef]
21. Feyerabend, F.; Fischer, J.; Holtz, J.; Witte, F.; Willumeit, R.; Drucker, H.; Vogt, C.; Hort, N. Evaluation of short-term effects of rare earth and other elements used in magnesium alloys on primary cells and cell lines. *Acta Biomater.* **2010**, *6*, 1834–1842. [CrossRef] [PubMed]
22. Hort, N.; Huang, Y.; Fechner, D.; Störmer, M.; Blawert, C.; Witte, F.; Vogt, C.; Drücker, H.; Willumeit, R.; Kainer, K.U.; et al. Magnesium alloys as implant materials—Principles of property design for Mg–RE alloys. *Acta Biomater.* **2010**, *6*, 1714–1725. [CrossRef] [PubMed]
23. Gao, L.; Chen, R.S.; Han, E.H. Effects of rare-earth elements Gd and Y on the solid solution strengthening of mg alloys. *J. Alloys Compd.* **2009**, *481*, 379–384. [CrossRef]
24. Peng, Q.; Wu, Y.; Fang, D.; Meng, J.; Wang, L. Microstructures and properties of melt-spun and as-cast Mg-20Gd binary alloy. *J. Rare Earths* **2006**, *24*, 466–470. [CrossRef]
25. Nodooshan, H.R.J.; Liu, W.C.; Wu, G.H.; Rao, Y.; Zhou, C.X.; He, S.P.; Ding, W.J.; Mahmudi, R. Effect of Gd content on microstructure and mechanical properties of Mg–Gd–Y–Zr alloys under peak-aged condition. *Mater. Sci. Eng. A* **2014**, *615*, 79–86. [CrossRef]
26. Xiaobo, Z.; Zhixin, B.; Zhangzhong, W.; Yujuan, W.; Yajun, X. Effect of LPSO structure on mechanical properties and corrosion behavior of as-extruded GZ51K magnesium alloy. *Mater. Lett.* **2016**, *163*, 250–253.
27. Myrissa, A.; Braeuer, S.; Martinelli, E.; Willumeit-Römer, R.; Goessler, W.; Weinberg, A.M. Gadolinium accumulation inorgans of Sprague-Dawley® rats after implantation of abiodegradable magnesium-gadolinium alloy. *Acta Biomater.* **2017**, *48*, 521. [CrossRef]
28. Xuenan, G.; Yufeng, Z.; Yan, C.; Shengping, Z.; Tingfei, X. In vitro corrosion and biocompatibility of binary magnesium alloys. *Biomaterials* **2009**, *30*, 484–498. [CrossRef]
29. Wang, X.; Dong, L.H.; Li, J.T.; Li, X.L.; Ma, X.L.; Zheng, Y.F. Microstructure, mechanical property and corrosion behavior of interpenetrating (HA+β-TCP)/MgCa composite fabricated by suction casting. *Mater. Sci. Eng. C* **2013**, *33*, 4266–4273. [CrossRef]
30. Salahshosor, M.; Li, C.; Liu, Z.Y.; Fang, X.Y.; Guo, Y.B. Surface integrity and corrosion performance of biomedical magnesiumcalcium alloy processed by hybrid dry cutting-finish burnishing. *J. Mech. Behav. Biomed. Mater.* **2018**, *78*, 246–253. [CrossRef]
31. Staiger, M.P.; Pietak, A.M.; Huadmai, J.; Dias, G. Magnesium and its alloys as orthopedic biomaterials: A review. *Biomaterials* **2006**, *27*, 1728–1734. [CrossRef] [PubMed]
32. Zeng, R.C.; Dietzel, W.; Witte, F.; Hort, N.; Blawert, C. Progress and challenge for magnesium alloys as biomaterials. *Adv. Eng. Mater.* **2008**, *10*, B3–B14. [CrossRef]
33. Li, Z.; Gu, X.; Lou, S.; Zheng, Y. The development of binary Mg-Ca alloys for use as biodegradable materials within bone. *Biomaterials* **2008**, *29*, 1329–1344. [CrossRef] [PubMed]
34. Atrens, A.; Song, G.L.; Liu, M.; Shi, Z.; Cao, F.; Dargusch, M.S. Review of recent developments in the field of magnesium corrosion. *Adv. Eng. Mater.* **2015**, *17*, 400–453. [CrossRef]
35. Yu, H.D.; Zhang, Z.Y.; Win, K.Y.; Chan, J.; Teoh, S.H.; Han, M.Y. Bioinspired fabrication of 3D hierarchical porous nanomicrostructures of calcium carbonate for bone regeneration. *Chem. Commun.* **2010**, *35*, 6578. [CrossRef] [PubMed]
36. Maradze, D.; Capel, A.; Martin, N.; Lewis, M.P.; Zheng, Y.; Liu, Y. In vitro investigation of cellular effects of magnesium and magnesium-calcium alloy corrosion products on skeletal muscle regeneration. *J. Mater. Sci. Technol.* **2019**, *35*, 2503–2512. [CrossRef]
37. Schwander, M.; Leu, M.; Stumm, M.; Dorchies, O.M.; Ruegg, U.T.; Schittny, J.; Müller, U. β1 Integrins Regulate Myoblast Fusion and Sarcomere Assembly. *Dev. Cell* **2003**, *4*, 673–685. [CrossRef]
38. Zeng, R.C.; Qi, W.C.; Cui, H.Z.; Zhang, F.; Li, S.Q.; Han, E.H. In vitro corrosion of as-extruded Mg–Ca alloys—The influence of Ca concentration. *Corros. Sci.* **2015**, *96*, 23–31. [CrossRef]
39. Salahshoor, M.; Guo, Y. Biodegradable Orthopedic Magnesium-Calcium (MgCa) Alloys, Processing, and Corrosion Performance. *Materials* **2012**, *5*, 135–155. [CrossRef]
40. Jeong, Y.S.; Kim, W.J. Enhancement of mechanical properties and corrosion resistance of Mg–Ca alloys through microstructural refinement by indirect extrusion. *Corros. Sci.* **2014**, *82*, 392–403. [CrossRef]
41. Master Alloys Supplier Website. Available online: http://www.hbnewmaterial.com/supplier-129192-master-alloy (accessed on 15 June 2020).
42. Lupescu, S.; Istrate, B.; Munteanu, C.; Minciuna, M.G.; Focsaneanu, S.; Earar, K. Characterization of Some Master Mg-X System (Ca, Mn, Zr, Y) Alloys Used in Medical Applications. *Rev. Chim.* **2017**, *68*, 1408–1413. [CrossRef]
43. Istrate, B.; Munteanu, C.; Lupescu, S.; Chelariu, R.; Vlad, M.D.; Vizureanu, P. Electrochemical Analysis and In Vitro Assay of Mg-0.5Ca-xY Biodegradable Alloys. *Materials* **2020**, *13*, 3082. [CrossRef] [PubMed]
44. Mosmann, T. Rapid colorimetric assay for cellular growth and survival: Application to proliferation and cytotoxicity assays. *J. Immunol. Methods* **1983**, *65*, 55–63. [CrossRef]
45. Vlad, M.D.; Valle, L.J.; Poeată, I.; Barracó, M.; López, J.; Torres, R.; Fernández, E. Injectable iron-modified apatitic bone cement intended for kyphoplasty: Cytocompatibility study. *J. Mater. Sci. Mater. Med.* **2008**, *19*, 3575–3583. [CrossRef]
46. Vlad, M.D.; Valle, L.J.; Poeată, I.; López, J.; Torres, R.; Barracó, M.; Fernández, E. Biphasic calcium sulfate dihydrate/iron-modified alpha-tricalcium phosphate bone cement for spinal applications: In vitro study. *Biomed. Mater.* **2010**, *5*, 025006. [CrossRef] [PubMed]

47. Liu, M.; Schmutz, P.; Uggowitzer, P.; Song, G.; Atrens, A. The influence of Y (Y) on the corrosion of Mg–Y binary alloys. *Corros. Sci.* **2010**, *52*, 3687–3701. [CrossRef]
48. Tong, X.; Zai, L.; You, G.; Wu, H.; Wen, H.; Long, S. Effects of bonding temperature on microstructure and mechanical properties of diffusion-bonded joints of as-cast Mg–Gd alloy. *Mater. Sci. Eng. A* **2019**, *767*, 138408. [CrossRef]
49. Kang, L.; Liang, Z.; Guohua, W.; Wencai, L.; Wenjian, D. Effect of Y and Gd content on the microstructure and mechanical properties of Mg–Y–RE alloys. *J. Magnes. Alloy.* **2019**, *7*, 345–354.
50. Junxiu, C.; Lili, T.; Xiaoming, Y.; Ke, Y. Effect of minor content of Gd on the mechanical and degradable properties of as-cast Mg-2Zn-xGd-0.5Zr alloys. *J. Mater. Sci. Technol.* **2019**, *35*, 503–511.
51. Sudholz, A.D.; Gusieva, K.; Chen, X.B.; Muddle, B.; Gibson, M.; Birbilis, N. Electrochemical behaviour and corrosion of Mg–Y alloys. *Corros. Sci.* **2011**, *53*, 2277–2282. [CrossRef]
52. Südholz, A.D.; Kirkland, N.T.; Buchheit, R.G.; Birbilis, N. Electrochemical Properties of Intermetallic Phases and Common Impurity Elements in Magnesium Alloys. *Electrochem. Solid-State Lett.* **2011**, *14*, C5–C7. [CrossRef]
53. Li, Z.; Sun, S.; Chen, M.; Fahlman, B.D.; Liu, D.; Bi, H. In vitro and in vivo corrosion, mechanical properties and biocompatibility evaluation of MgF2-coated Mg-Zn-Zr alloy as cancellous screws. *Mater. Sci. Eng. C* **2017**, *75*, 1268–1280. [CrossRef] [PubMed]
54. Gilles, R.; Belkhir, M.; Compere, P.; Libioulle, C.; Thiry, M. Effect of high osmolarity acclimation on tolerance to hyperosmotic shocks in L929 cultured cells. *Tissue Cell* **1995**, *27*, 679–687. [CrossRef]
55. Esmaily, M.; Svensson, J.E.; Fajardo, S.; Birbilis, N.; Frankel, G.S.; Virtanen, S.; Arrabal, R.; Thomas, S.; Johansson, L.G. Fundamentals and advances in magnesium alloy corrosion. *Prog. Mater. Sci.* **2017**, *89*, 92–193. [CrossRef]
56. ISO 10993-5:2009—Biological Evaluation of Medical Devices—Part 5: Tests for In Vitro Cytotoxicity. Available online: http://nhiso.com/wp-content/uploads/2018/05/ISO-10993-5-2009.pdf (accessed on 25 September 2020).
57. Mbele, G.O.; Deloulme, J.C.; Gentil, B.J.; Delphin, C.; Ferro, M.; Garin, J.; Takahashi, M.; Baudier, J. The zinc and calcium-binding S100B interacts and co-localizes with IQGAP1 during dynamic rearrangement of cell membranes. *J. Biol. Chem.* **2002**, *277*, 49998–50007. [CrossRef] [PubMed]
58. Yang, L.; Hort, N.; Laipple, D.; Höche, D.; Huang, Y.; Kainer, K.U.; Willumeit, R.; Feyerabend, F. Element distribution in the corrosion layer and cytotoxicity of alloy Mg–10Dy during in vitro biodegradation. *Acta Biomater.* **2013**, *9*, 8475–8487. [CrossRef] [PubMed]

Article

Laser Induced Method to Produce Curcuminoid-Silanol Thin Films for Transdermal Patches Using Irradiation of Turmeric Target

Alexandru Cocean [1], Iuliana Cocean [1], Nicanor Cimpoesu [2,*], Georgiana Cocean [3], Ramona Cimpoesu [2], Cristina Postolachi [1], Vasilica Popescu [4] and Silviu Gurlui [1,*]

1. Faculty of Physics, Atmosphere Optics, Spectroscopy and Laser Laboratory (LOASL), Alexandru Ioan Cuza University of Iasi, 11 Carol I Boulevard, 700506 Iasi, Romania; alexcocean@yahoo.com (A.C.); iuliacocean@gmail.com (I.C.); tina.postolaki@gmail.com (C.P.)
2. Faculty of Material Science and Engineering, Gheorghe Asachi Technical University of Iasi, 59A Mangeron Boulevard, 700050 Iasi, Romania; ramona.cimpoesu@tuiasi.ro
3. Rehabilitation Hospital Borsa, 1 Floare de Colt street, 435200 Borsa, Romania; cocean.georgiana@yahoo.com
4. Department of Chemical Engineering in Textiles and Leather, Gheorghe Asachi Technical University of Iasi, 29 Boulevard Dimitrie Mangeron, 700050 Iasi, Romania; vpopescu65@yahoo.com
* Correspondence: nicanor.cimpoesu@tuiasi.ro (N.C.); sgurlui@uaic.ro (S.G.)

Abstract: A new possible method to produce a transdermal patch is proposed in this paper. The study refers to the pulsed laser deposition method (PLD) applied on turmeric target in order to obtain thin layers. Under high power laser irradiation of 532 nm wavelength, thin films containing curcuminoids were obtained on different substrates such as glass and quartz (laboratory investigation) and hemp fabric (practical application). Compared FTIR, SEM-EDS and LIF analyses proved that the obtained thin film chemical composition is mainly demethoxycurcumin and bisdemethoxycurcumin which is evidence that most of the curcumin from turmeric has been demethixylated during laser ablation. Silanol groups with known role into dermal reconstruction are evidenced in both turmeric target and curcuminoid thin films. UV–VIS reflection spectra show the same characteristics for all the curcuminoid thin films, indicating that the method is reproducible. The method proves to be successful for producing a composite material, namely curcuminoid transdermal patch with silanol groups, using directly turmeric as target in the thin film deposited by pulsed laser technique. Double layered patch curcuminoid—silver was produced under this study, proving compatibility between the two deposited layers. The silver layer added on curcuminoid-silanol layer aimed to increase antiseptic properties to the transdermal patch.

Keywords: PLD; turmeric; curcuminoid-silanol films; transdermal patch; demetoxilation; SEM-EDS; LIF; hemp composite

Citation: Cocean, A.; Cocean, I.; Cimpoesu, N.; Cocean, G.; Cimpoesu, R.; Postolachi, C.; Popescu, V.; Gurlui, S. Laser Induced Method to Produce Curcuminoid-Silanol Thin Films for Transdermal Patches Using Irradiation of Turmeric Target. *Appl. Sci.* **2021**, *11*, 4030. https://doi.org/10.3390/app11094030

Academic Editor: Andrea Li Bassi

Received: 29 March 2021
Accepted: 25 April 2021
Published: 28 April 2021

Publisher's Note: MDPI stays neutral with regard to jurisdictional claims in published maps and institutional affiliations.

Copyright: © 2021 by the authors. Licensee MDPI, Basel, Switzerland. This article is an open access article distributed under the terms and conditions of the Creative Commons Attribution (CC BY) license (https://creativecommons.org/licenses/by/4.0/).

1. Introduction

Turmeric effects on human's health have been intensively studied lately, after its benefits had been observed during the long time use as spice in food. Extracted from turmeric, curcumin, along with the other curcuminoids, were found as carriers of curative effects for a long list of diseases such as anti-inflammatory including in rheumatoid arthritis, antioxidant, increase of brain-derived neurotrophic factor (BDNF), benefits against depression, [1,2]. Cyto-protective, anticancer properties and immunomodulating effect of turmeric have been also proven [3,4] and its antimicrobial, antioxidant, and astringents properties also recommend it for stomatology and generally, oral uses [5,6]. A turmeric protective effect against natural and chemically-induced toxicity has been also reported [7]. Studies showed favorable piperine influences on pharmacokinetics of curcumin [8]; otherwise, curcumin would not be properly absorbed. However, turmeric itself contains piperine that allows a better absorption into the body, including dermal absorption. Thus, turmeric should be preferred as the basic natural component when to start a new procedure for obtaining derivatives to be used as pharmaceutical products to harness turmeric entirely potential, though the curcumin percentage is said to be low in turmeric. Sometimes, the natural composite components would interact to provide a synergic effect for a highly efficiency in order to obtain the desired purpose.

Hybrid layered double hydroxides-curcumin thin films were obtained through MAPLE (Assisted Pulsed Laser Evaporation) for medical purpose [9] and very encouraging results on the incorporation of curcumin in liquid nano-domains embedded into polymeric films for dermal application were realized [10]. The disadvantage of these methods is that they require the prior extraction of curcumin from turmeric which is difficult and low yield when performed by conventional methods due to the low solubility of curcumin in various solvents [11,12]. The method proposed by this study aims to solve the problem of poor solubility of turmeric in various solvents that makes it less available to pharmaceutical purposes [12]. The novelty of the study presented herein consists in use of the technique PLD (Pulsed Laser Deposition) to ablate turmeric and deposit on different materials with the further goal to produce medical devices, mainly as medical transdermal patches. The PLD technique applied directly to the turmeric target is an absolutely new method of extracting curcuminoids from turmeric. The final purpose is to transfer the curcuminoids obtained in the deposited thin layer through skin into the body and/or to heal wounds, dermatitis, to regenerate and prevent infections, silanol groups from turmeric contributing as well to such effects along with the other components. Silanol benefits for the dermal cure have been already proven [13] and have been successfully used in cosmetics.

Double layered patch curcuminoid—silver was also produced under this study and proved compatibility between constituents, with the aim to increase antiseptic properties. Further studies in medical environment of an experimental patch are proposed to several particular clinics and the results will be evaluated in a different article.

2. Materials and Methods

The turmeric powder is of Indian provenience, commercialized in Romania by Sanflora Bucuresti, Romania, with determined humidity of 9.32% and with an average granular size of the order of tens of micrometers, as can be seen from the images obtained by optical microscopy of the turmeric grains compared to a grid of 0.1 mm × 0.1 mm (Figure 1a). From the same optical microscopy images, the tendency of agglomeration and/or aggregation of turmeric powder grains is visible by the presence of particles of different shapes and sizes that reach hundreds of micrometers. However, from the point of view of the procedure used in laser ablation, this is not an impediment, because the powder is to be pressed in order to obtain the target. In this regard, the turmeric powder was pressed using a pressure of 100 atm in a ring-shaped stainless steel mold to form the target. The pressure of 100 atm is enough to compact the powder but should not affect the turmeric properties because when removed from the mold, the resulted powder presents grains of similar size to the bigger grains in the initial turmeric powder, probably due to agglomeration/aggregation during compression (Figure 1b). This means that the physical properties have not undergone any changes or only minor changes, namely related to the aggregation of the grains. Moreover, the pressure was low enough not to induce chemical changes in the compressed turmeric powder. However, the turmeric powder was compressed by the same method, mixed with potassium bromide when the FTIR analysis was performed, the resulting spectra describing exactly the target used for PLD. Images from the preparation of the turmeric powder target by pressing, the positioning of the target in the vacuum chamber, together with images during the pulsed laser deposition process are presented in Figure 2a–g.

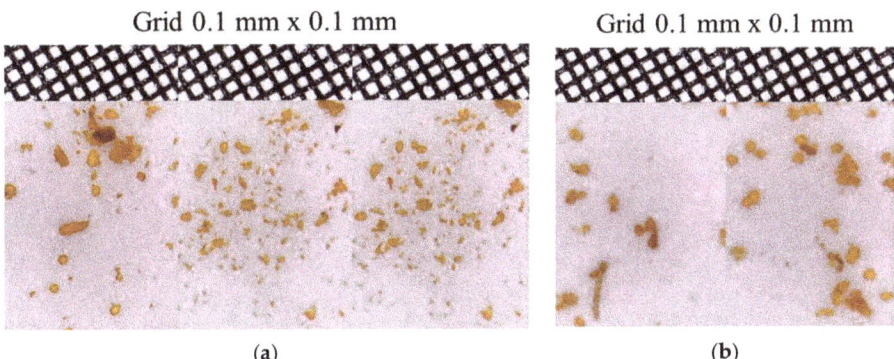

Figure 1. Optical microscope images of Turmeric powder grains compared with a grid scale of 0.1 mm × 0.1 mm before compression in the mold (**a**) and after compression in the mold (**b**).

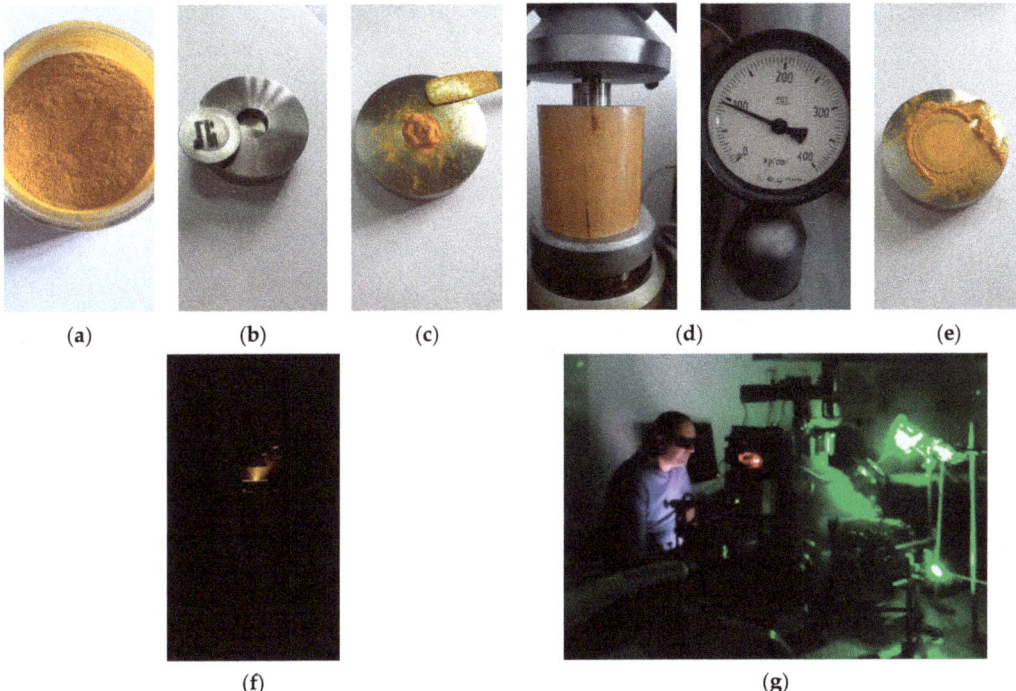

Figure 2. Images of materials and means used to produce the turmeric target and its displacement into the work system: Turmeric powder (**a**), ring-shaped stainless steel mold (**b**), filling the mold with turmeric powder (**c**), pressing the turmeric powder into the mold at 100 atm (**d**), turmeric target formed into the stainless steel ring (**e**), target placed into the vacuum chamber and target irradiation (**f**), images during work with laser beam irradiation at 532 nm wavelength (**g**).

The resulted target in size of 10 mm radius and 4 mm height was irradiated with 10 ns pulsed laser beam of 336 µm spot radius and 532 nm wavelength with 10 Hz repetition rate, using a YG 981E/IR-10 laser system at 25 J/cm^2 fluence for 30 min, which results in 18,000 pulses [14,15]. Distance between target and support was of 2.5 cm and the pressure in the deposition chamber 10^{-2} Torr. During deposition, the target is moved by a software-assisted mechanism. Due to the circular shape of the target, in this case the spiral movement was selected with the starting point in the center of the target. The equipment used is in Atmosphere Optics, Spectroscopy and Lasers Laboratory (LOASL) and it is presented schematically in Figure 3a. The wavelength of 532 nm is used for turmeric ablation and deposition as it has been previously proven that it preserves the organic structure better with minimum bonds breakage [14].

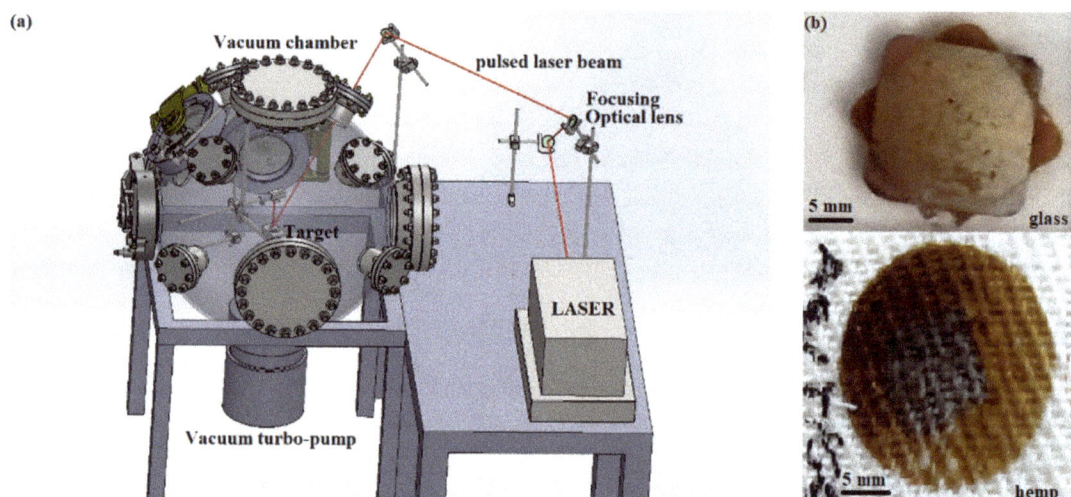

Figure 3. Pulsed laser deposition of Turmeric: (**a**) schematic 3D view of PLD equipment (Atmosphere Optics, Spectroscopy and Lasers Laboratory (LOASL), Faculty of Physics from Alexandru Ioan Cuza University of Iasi); (**b**) depositions of turmeric and silver on glass and hemp fabric.

Pulsed laser depositions were performed on glass and quartz slabs (in order to obtain experimental results), as well as on hemp twill fabric (for a possible medical application), Figure 1b. The hemp fabric material was used as substrate in order to observe the practical possibility of growing new thin layers on this type of complex surface. Known antimicrobial and antifungal silver properties, a double layer was also produced aimed to enhance the healing and antiseptic properties. Turmeric was first deposited and then silver thin film on top of it. This way, five samples were obtained: turmeric on glass slab (T/GLS), turmeric on quartz slab (T/QTZ), turmeric on hemp twill fabric (T/HMP), silver on turmeric coating on glass slab (Ag/T/GLS) and silver on turmeric coating on hemp twill fabric (Ag/T/HMP). Images of the material depositions (experimental glass and hemp substrates) were realized using scanning electron microscopy (SEM: VegaTescan LMH II, SE detector, 30 kV) in order to obtain details at higher amplification. Energy dispersive spectroscopy (EDS, Bruker X-flash) was used to identify the chemical elements from the thin layers obtained through PLD.

Fourier transform infrared spectroscopy (FTIR) performed with Versatile FT-IR Laboratory Spectrometer MB3000 on dry powder samples incorporated in potassium bromide provided information on the functional groups of chemical compounds deposited. The information on the chemical composition was completed by LIF spectrometry that was conducted using the UV laser beam of 355 nm wavelength of the same laser system YG 981E/IR-10. For data acquisition, a high resolution spectrometer (750 mm focal length) Acton 2750i coupled with Roper Scientific PIMAX3 ICCD camera, 1024 × 1024 pixels, 2 ns minimum gate time was used.

3. Results

The thickness of the curcuminoid layer produced (Curcuminoid-silanol film/Quartz) measured with Stylus Profiler DektakXT Bruker was found as being of four micrometers on an area of 226.8 mm^2. As this experiment was designed to evaluate quality of deposition and to set-up the parameters and conditions for ablation, the deposition efficiency was not calculated due to the difficulties of accurate measurement of the ablated material on the target. However, given the fine "trace" left on the target by laser radiation due to ablation, compared to the deposition surface and the thickness of the layer, it is expected that the deposition yield with the PLD method will be very high.

A sample of curcuminoid powder was collected scraping the thin film obtained by PLD technique using same procedure as reported before by Cocean et al. [14] and it was used for FTIR analysis, resulting the spectrum of *Curcuminoid-silanol fil*. FTIR analysis was also performed on turmeric powder used as target, resulting the spectrum of *Turmeric*, as well as on pure curcumin powder, namely the spectrum of *Curcumin*. All the three spectra were compared. 5. Curcuminoid-silanol thin films fluorescence was analyzed compared to the turmeric powder using laser induced fluorescence (LIF.. Chemical analysis was completed by elemental composition performed by SEM-EDS spectroscopy when also morphological information was obtained. UV–Vis reflection spectra provided information regarding optical properties of the obtained curcuminoid films *Curcuminoid-silanol film/Glass*, *Curcuminoid-silanol film/Quartz*, *Curcuminoid-silanol film/HMP*, *Ag/Curcuminoid-silanol film/Glass*, and *Ag/Curcuminoid-silanol film/Hemp*.

3.1. Fourier Transform Infrared Spectroscopy (FTIR) Analysis

For the deposition with PLD method, the turmeric powder compressed into a disk shape target without any additives was used. That provides information on the starting chemical composition where the specific groups in curcuminoids are expected to be found, such as shown in the chemical formulae in Figure 4. Fourier transform infrared spectroscopy (FTIR) analysis of both target and obtained thin film resulted into spectra of high similitude (Figure 5), proving that curcuminoid-silanol films were obtained.

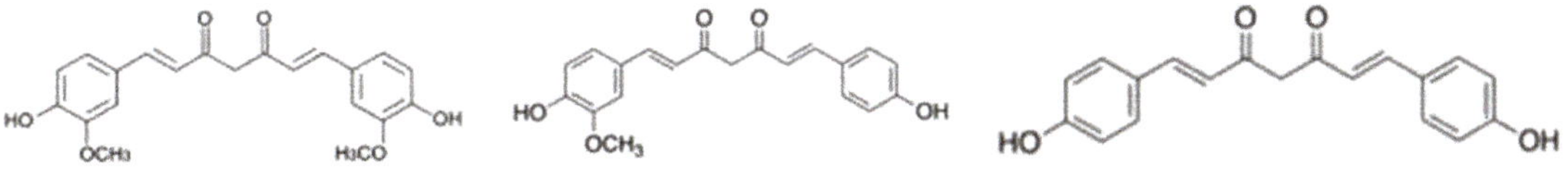

Curcumin Demethoxycurcumin Bisdemethoxycurcumin

Figure 4. Chemical structures of curcuminoids.

In Figure 5, for the *Turmeric* FTIR spectrum, predominant curcumin component is observed, while for the thin film obtained by PLD (*Curcuminoid-silanol film* spectrum), demethoxylation is evidenced by the drastic mitigation in transmittance intensity, which becomes barely visible, of the 2925 cm^{-1} and 2853 cm^{-1} bands specific for C–H aliphatic stretching of the acetal (methoxy) groups O–CH$_3$, indicating increase in the other two curcuminoid components (demethoxycurcumin, and bisdemethoxycurcumin). Demethoxylation is also denoted by lower intensity of bands in the thin film spectrum compared with the turmeric powder, bands corresponding to specific groups. Therefore, intensity mitigation is noticed for the 1279 cm^{-1} band of the (ar)C–O–(al)C groups stretching asymmetric, usually vibrating in the range (1275–1200 cm^{-1}) [16], in the same range with (ar)C–H in plane deformation vibration (1250–950 cm^{-1}) while the (ar)C–C group expected in the range 3080–3030 cm^{-1} [16] overlaps with the broad peak at 3440 cm^{-1} in turmeric spectrum and it is very weak in the thin film spectrum. Moreover, a decrease of the intensity of the bands is observed for those at 1079 cm^{-1} and 1024 cm^{-1} in the thin film spectrum compared with the turmeric powder spectrum corresponding to (ar)C–O–(al)C stretching symmetric vibrations, usually occurring in the range of 1075–1020 cm^{-1} [16]. In the same range are the specific silanol terminal groups as it is presented below. The other specific vibrations for the curcuminoids are observed in both FTIR spectra of turmeric powder and deposited thin film.

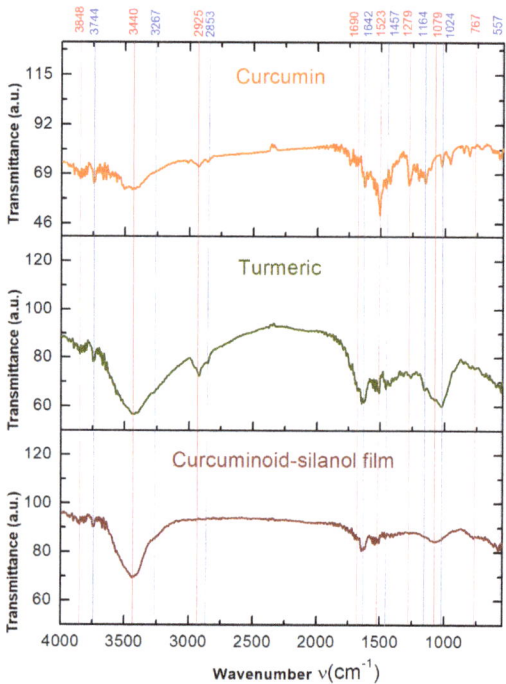

Figure 5. FTIR spectra of curcumin powder (*Curcumin*), turmeric powder (*Turmeric*), *Curcuminoid-silanol film* (thin film obtained by PLD technic)

It is important to notice silanol groups which are identified in both turmeric powder and thin film. Thermal resistivity of the silanol groups explains their preservation during ablation and occurrence in the deposited thin films. The band at 3848 cm^{-1} is of O–H free stretching vibrations [17] that could be associated to Si–OH groups [18] and the 3744 cm^{-1} band has been previously proven and reported for the terminal silanol groups [19–21], the IR spectroscopy generally assigning the bands in the range 3700–3200 cm^{-1} to Si–OH stretching vibrations [18]. The bands at 1457 cm^{-1} and 1279 cm^{-1} are of the (Si–)CH$_3$ group asymmetric and symmetric, respectively, deformation vibrations [16]. The 1024 cm^{-1} band of the Si–OH deformation vibrations (~1030 cm^{-1}), but also 1079 cm^{-1} of Si–O–Si stretching vibrations (1090–1030 cm^{-1}) [16] are lower in intensity for the thin film than in the turmeric powder spectrum, showing that some of the silanol terminal groups are removed during ablation process. Silanol terminal groups are also evidenced with the 767 cm^{-1} band assessed to Si–C stretching vibration in the 850–650 cm^{-1} range and to (Si–)CH$_3$ skeletal vibration that usually occurs at ~765 cm^{-1}) [16].

The O–H free and H-bonded phenolic groups are evidenced by the stretching vibrations at 3440 cm^{-1} [16]. The 3267 cm^{-1} band indicates C–H aromatic stretching vibrations [16], aromatic character being also evidenced in the C=C aromatic skeletal vibrations of 1523 cm^{-1} band [16]. The 2925 cm^{-1} band is assessed to C–H aliphatic stretching [16] together with the C–H stretching vibration in O–CH$_3$ acetals denoted by the band at 2853 cm^{-1} [14]. The 1690 cm^{-1} band is assessed to C=O stretching, aliphatic, attached to aromatic groups that causes a shift to lower wavenumbers from ~1715 to 1695–1695 cm^{-1}

(wavenumber decreases with increasing ring size) [16], a band at 1642 cm^{-1} of C=O stretching vibrations being also present [17]. The band at 557 cm^{-1} indicates C–S groups [16].

3.2. Scanning Electron Microscopy with Energy Dispersive Spectroscopy (SEM-EDS)

SEM analysis provides information on the morphology (Figure 6) and elemental composition (Table 1) of the turmeric powder (*Turmeric*) and the thin films obtained by PLD technic (*Curcuminoid-silanol film*).

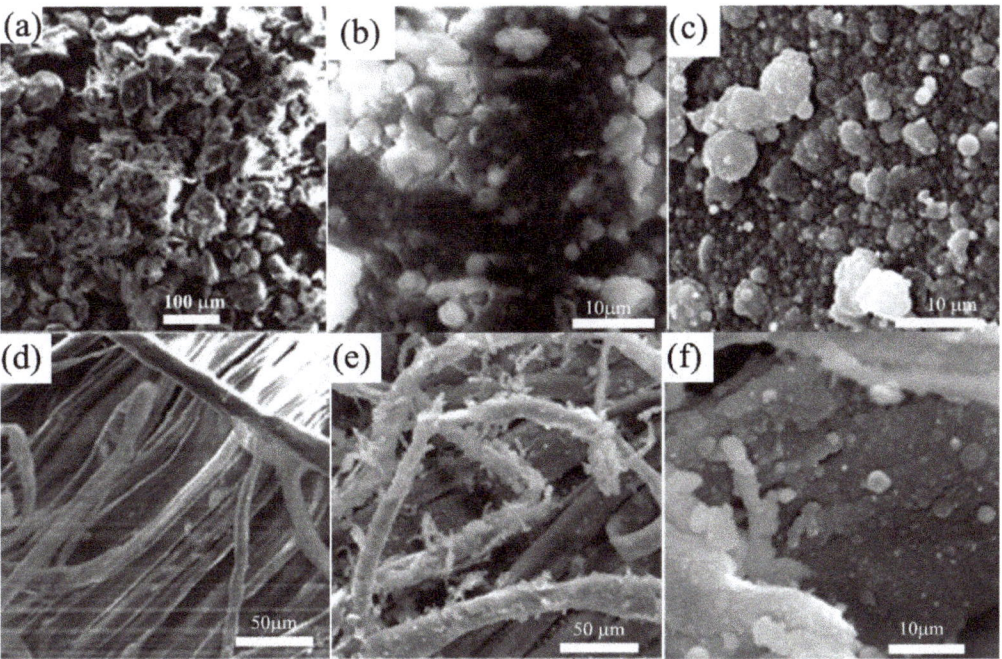

Figure 6. SEM images of turmeric powder (Turmeric) 350× (**a**); Curcuminoid-silanol film/Glass 5kx (**b**); Ag/Curcuminoid-silanol film/Glass 5kx (**c**); Curcuminoid-silanol film/Hemp 1kx (**d**); Ag/Curcuminoid-silanol film/Hemp 1kx (**e**); detail of Ag/Curcuminoid-silanol film/Hemp 5kx (**f**).

Table 1. Compared elemental composition of turmeric powder and thin films deposited on glass slab and on hemp fabric, respectively.

	Atom (%)			Weight (%)			
Element	Turmeric Powder	Curcuminoid-Silanol Film/Glass	Curcuminoid-Silanol Film/Hemp	Turmeric Powder	Curcuminoid-Silanol Film/Glass	Curcuminoid-Silanol Film/Hemp	EDS Error %
Carbon	63.89	65.39	65.96	56.49	51.84	53.37	2.2
Oxygen	35.49	29.83	33.43	41.80	31.50	39.41	1.2
Iron	0.01	1.62	0.29	0.02	5.96	0.29	0.01
Copper	-	0.99	0.01	-	4.18	0.04	0.09
Potassium	0.53	1.17	0.15	1.52	3.01	0.43	0.11
Chromium	0.01	0.46	0.09	0.01	1.60	0.33	0.1
Nickel	-	0.31	0.03	-	1.19	0.11	0.13
Zinc	-	0.12	0.01	-	0.51	0.01	0.05
Silicon	0.08	0.11	0.04	0.17	0.21	0.09	0.05

Standard deviations: C: ±1.2, O: ±0.8, Si: ±0.01, Fe: ±0.001, Cu: ±0.001, K: ±0.001, Cr: ±0.005, Ni: ±0.01 and Zn: ±0.001.

SEM images show polymeric agglomerations for the *Curcuminoid-silanol film/Glass* and *Curcuminoid-silanol film/Hemp* proving self-organization properties, and a good coating of the fibers in the *Curcuminoid-silanol film/Hemp*. Added silver particles are uniformly dispersed on top of the curcuminoid thin films and within the hemp fibers from the fabric texture. The dimensions of measured silver particles vary from 2 to 3 µm droplets to a nano-metric layer deposited on the substrate [22]. The second role of silver layer is to fix and sustain the curcuminoid-silanol thin film on the substrate. Both variants with and without very thin silver layer present interests in medical field.

Elemental composition shows, among specific organic elements such as carbon and oxygen, an atomic percentage of 0.08% silicon in turmeric powder and 0.04% in the composition of the curcuminoid thin film deposited on hemp. The 0.11% of silicon determined in the *Curcuminoid-silanol film/Glass* could be due to the influence of the substrate composition. Carbon elemental composition is similar in turmeric powder and thin films deposited on both glass and hemp, with only slight differences. In atomic percentage, the variation of carbon percentage is from 63.89% in turmeric powder to 65.38% in *Curcuminoid-silanol film/Glass* and 65.95% in *Curcuminoid-silanol film/Hemp*, the latter being under the influence of the hemp substrate composition. The atomic composition of oxygen was of 35.49% for turmeric powder, 29.83% for *Curcuminoid-silanol film/Glass*, and 33.43% for *Curcuminoid-silanol film/Hemp*.

The balance to 100% is split between different other elements such as Fe, Cu, K, Cr, Zn, and they reflect the vegetal material property to absorb metals, the plant roots and rhizomes retaining most of them. The same mechanism made possible absorption of silicon which proves affinity for the curcuminoids to form silanol bond on their terminal groups.

Laser induced fluorescence (LIF). LIF also provides information about chemical structure of the investigated material when fluorophore groups are present and about UV laser beam interaction with the material, which are the curcuminoids in this case. Laser induced fluorescence using 355 nm laser beam wavelength evidenced fluorescence spectra as shown in Figure 7. The sharp peaks at about 514 nm and about 550 nm may be assessed to the sulfur [23] and silanol groups, respectively [24] in the curcuminoid molecules. The latter one was observed for the silanol alkyl groups. Silanol groups have been identified in the FTIR spectra of Figure 5 as terminal groups. That, together with the LIF spectra, shows that some of the silanol groups are bond to the methoxy groups of curcumin, explaining the mitigation in FTIR spectrum intensity of the thin film for both silanol and methoxy. The information from the spectra of Figure 5 is that a partial demethoxylation of curcumin takes place, methoxy groups leaving the molecule together with the silanol groups bond to methoxy. As silanol groups are alkyl ones, their occurrence in the thin film spectrum is an evidence that not all the methoxy groups are removed during ablation and the thin film is a mixture of all the three curcuminoids (curcumin, demethoxycurcumin, and bisdemethoxycurcumin) characteristic to turmeric, just their ratio is changed.

The high ratio of curcumin in the target became low in the thin curcuminoid film, where demethoxycurcumin and bisdemethoxycurcumin predominate.

Specific 562 nm fluorescent emission for turmeric under laser irradiation of 355 nm is lower by two orders (100 times) in intensity for the curcuminoid thin films compared to turmeric powder.

Two tower like structures emissions in 481–508 nm range and 588–627 nm range in all LIF spectra of Figure 7 evidence chemical reactions [25] including radical formation, mainly OH radicals [14]. Secondary peaks are also evidenced at 426 nm, 454 nm, 667 nm, and 684 nm as Swan peaks characteristic to the carbon stars spectra, comets and to burning hydrocarbon fuels [26,27] that here could be from the organic carbon burning during ablation.

Silver thin film deposited on the curcuminoid films produce mitigation in the fluorescent emission due to the shielding effect on the absorption and emission, but also to the interference with the fluorophore groups of curcuminoids, especially C = O with high affinity for metals.

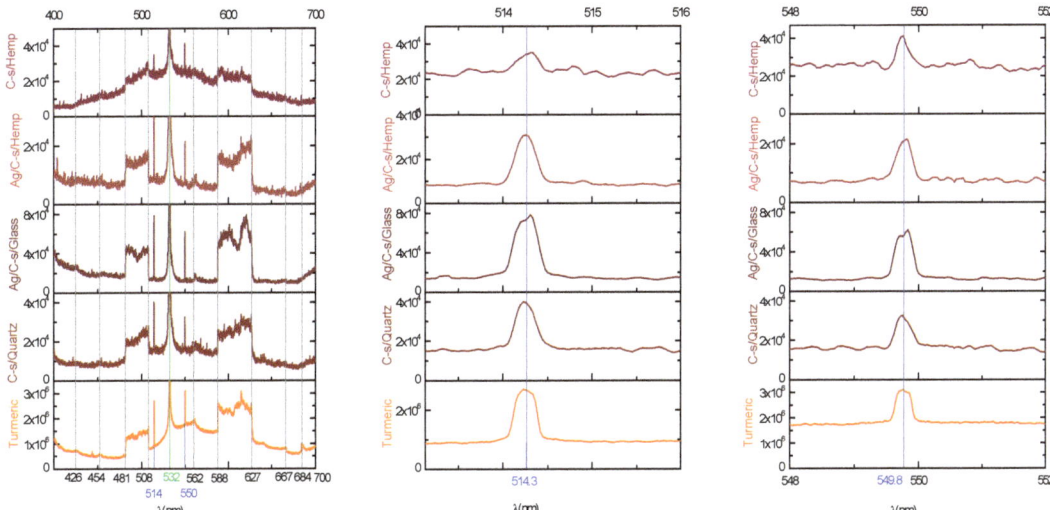

Figure 7. LIF compared spectra of Turmeric, Curcuminoid-silanol film/Quartz (*C-s/Quartz*), and Curcuminoid-silanol film/Hemp (*C-s/Hemp*) and the double layers Ag/Curcuminoid-silanol film/Glass (*Ag/C-s/Glass*) and Ag/Curcuminoid-silanol film/Hemp (*Ag/C-s/Hemp*).

3.3. UV–VIS Spectra of Thin Films

In Figure 8, reflection of the 578 nm, 664 nm, and 696 nm light components are noticed. A sharp peak at 436.7 nm is also noticed coming out from the main spectra that could be assessed to the UV lamp irradiation. All the deposited thin films reflect at the same wavelengths and very close intensities (Table 2).

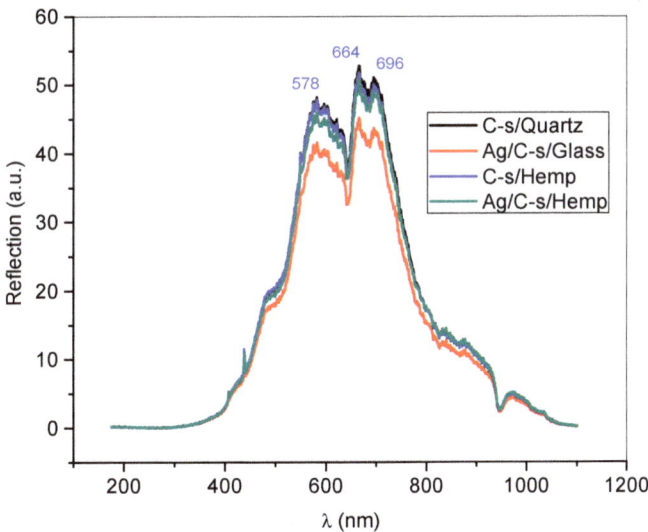

Figure 8. UV–VIS spectra of the thin films reflection properties: Curcuminoid-silanol film/Quartz (*C-s/Quartz*) and Curcuminoid-silanol film/Hemp (*C-s/Hemp*) and the double layers Ag/Curcuminoid-silanol film/Glass (*Ag/C-s/Glass*) and Ag/Curcuminoid-silanol film/Hemp (*Ag/C-s/Hemp*).

Table 2. Reflection intensities of the thin films deposited.

	Reflection Intensity (a.u.)		
	Λ = 578 nm	Λ = 664 nm	Λ = 696 nm
Curcuminoid-silanol film/Quartz	48.26	52.6	51.2
Curcuminoid-silanol film/Hemp	47.8	52.3	50.3
Ag/Curcuminoid-silanol film/Glass	45.7	51.2	48.9
Ag/Curcuminoid-silanol film/Hemp	41.1	45	43.6

The reflection intensity of the curcuminoid layers deposited on hemp decreases only by 0.95% at 578 nm, 0.57% at 664 nm, and 1.75% at 696 nm compared to curcuminoid films deposited on glass. In the case of depositing the additional silver layer, the decrease in intensity of depositions on hemp substrate compared to depositions on glass substrate is 10% at 578 nm, 12% at 664 nm, and 10.8% at 696 nm. In the case of comparison according to the substrate used, there is a decrease of only 4.3% at 578 nm, 2.1% at 664 nm, and 2.78% at 696 nm when the films are deposited on hemp, while when the films are deposited on glass, the differences are 14.83% at 578 nm, 14.4% at 664 nm, and 14.8% at 696 nm. It is observed that the smallest differences are for the single layers of curcuminoids deposited on different substrate (on glass and on hemp), as well as for the double layers of silver and curcuminoids deposited on hemp substrate. The largest variations in reflection intensity are recorded between the deposition of the curcuminoids layer and the deposition of the double layer of silver and curcuminoids on glass substrate. These observations show that the deposition of thin layers of curcuminoids starting from the turmeric target are reproducible with a margin of error below 2%. The increase in the variation of the reflection intensity after the deposition of the second layer, the silver one over the thin layer of curcuminoids, is explained due to the tendency to form droplets. In the case of hemp substrate, the smaller variation than in the case of glass substrate is due to the dispersion of nanoparticles that form the curcuminoid film and the silver and curcuminoid double film among the fibers and microfibrils of the fabric, as seen in electronic microscopy images of Figure 6e,f.

4. Conclusions

New method and new composite materials are reported with this paper. Starting from a natural material, turmeric powder, used as target in the pulsed laser deposition process, the thin films obtained prove to be suitable for medical purposes due to their composition in curcuminoids whose health benefits have been proven. It is also assumed, based on preliminary evaluation, that the method is reproducible which may recommend it for industrial purposes. The method is proven to be a possible solution to the low solubility of turmeric and its components when desired to obtain curcuminoid films or coatings on different substrates. Medical patch for dermal transfer of curative substances containing curcuminoids and silanol can be produced using the method as described herein. Double layered patch curcuminoid – silver is aimed to enhance antiseptic effect of the device. Medical analysis and tests are further required, dosing the quantity of curcuminoid layer deposited needed to provide an optimal transfer over a period of time and for specific diseases.

Author Contributions: Conceptualization, A.C., G.C., C.P., I.C., and S.G.; methodology, I.C., S.G., and A.C.; software, A.C.; validation, A.C., I.C., and S.G.; formal analysis, A.C., C.P., I.C., S.G., N.C., R.C., and V.P.; investigation, A.C., I.C., C.P., G.C., N.C., and S.G.; resources, S.G., N.C., V.P., and I.C.; data curation, A.C., I.C., and S.G.; writing—original draft preparation, I.C., S.G., and A.C.; writing—review and editing, I.C., N.C., R.C., and S.G.; visualization, A.C., S.G., and I.C.; supervision, S.G.; project administration, S.G.; funding acquisition, S.G. and N.C. All authors have read and agreed to the published version of the manuscript.

Funding: This research was funded by Ministry of Research, Innovation and Digitization, project FAIR_09/24.11.2020 and by the Executive Agency for Higher Education, Research, Development, and Innovation, UEFISCDI, ROBIM- project number PN-III-P4-ID-PCE2020-0332.

Institutional Review Board Statement: Not applicable.

Informed Consent Statement: Not applicable.

Conflicts of Interest: The authors declare no conflict of interest.

References

1. Kumar, A.; Singh, A.K.; Kaushik, M.S.; Mishra, S.K.; Raj, P.; Singh, P.K.; Pandey, K.D. Interaction of turmeric (*Curcuma longa* L.) with beneficial microbes: A review. *3 Biotech* **2017**, *7*, 357. [CrossRef]
2. Niranjan, A.; Prakash, D. Chemical constituents and biological activities of turmeric (*Curcuma longa* L.)—A review. *J. Food Sci. Technol.* **2008**, *45*, 109–116.
3. Subtaweesin, C.; Woraharn, W.; Taokaew, S.; Chiaoprakobkij, N.; Sereemaspun, A.; Phisalaphong, M. Characteristics of Curcumin-Loaded Bacterial Cellulose Films and Anticancer Properties against Malignant Melanoma Skin Cancer Cells. *Appl. Sci.* **2018**, *8*, 1188. [CrossRef]
4. Abu-Rizq, H.A.; Mansour, M.H.; Safer, A.M.; Afzal, M. Cyto-protective and immunomodulating effect of *Curcuma longa* in Wistar rats subjected to carbon tetrachloride-induced oxidative stress. *Inflammopharmacology* **2008**, *16*, 87–95. [CrossRef] [PubMed]
5. Alok, A.; Singh, I.D.; Singh, S.; Kishore, M.; Jha, P.C. Curcumin—Pharmacological Actions and its Role in Oral Submucous Fibrosis: A Review. *J. Clin. Diagn. Res.* **2015**, *9*, ZE01–ZE03. [CrossRef] [PubMed]
6. Sood, S.; Nagpal, M. Role of curcumin in systemic and oral health: An overview. *J. Nat. Sci. Biol. Med.* **2013**, *4*, 3–7. [CrossRef]
7. Hosseini, A.; Hosseinzadeh, H. Antidotal or protective effects of *Curcuma longa* (turmeric) and its active ingredient, curcumin, against natural and chemical toxicities: A review. *Biomed. Pharmacother.* **2018**, *99*, 411–421. [CrossRef]
8. Shoba, G.; Joy, D.; Joseph, T.; Majeed, M.; Rajendran, R.; Srinivas, P.S.S.R. Influence of Piperine on the Pharmacokinetics of Curcumin in Animals and Human Volunteers. *Planta Med.* **1998**, *64*, 353–356. [CrossRef] [PubMed]
9. Andrei, F.; Vlad, A.; Birjega, R.; Tozar, T.; Secuc, M.; Urzica, I.; Dinescu, M.; Zavoianu, R. Hybrid layered double hydroxides-curcumin thin films deposited via Matrix Assisted Pulsed Laser Evaporation-MAPLE with photoluminescence properties. *Appl. Surf. Sci.* **2019**, *478*, 754–761. [CrossRef]
10. Abramov, E.; Garti, N. Incorporation of curcumin in liquid nanodomains embedded into polymeric films for dermal application. *Colloids Surf. B Biointerfaces* **2021**, *198*, 111468. [CrossRef]
11. Carvalho, D.D.M.; Takeuchi, K.P.; Geraldine, R.M.; De Moura, C.J.; Torres, M.C.L. Production, solubility and antioxidant activity of curcumin nanosuspension. *Food Sci. Technol.* **2015**, *35*, 115–119. [CrossRef]
12. Suresh, K.; Nangia, A. Curcumin: Pharmaceutical solids as a platform to improve solubility and bioavailability. *CrystEngComm* **2018**, *20*, 3277–3296. [CrossRef]
13. Herreros, F.O.C.; Cintra, M.L.; Adam, R.L.; De Moraes, A.M.; Metze, K. Remodeling of the human dermis after application of salicylate silanol. *Arch. Dermatol. Res.* **2007**, *299*, 41–45. [CrossRef]
14. Cocean, I.; Cocean, A.; Postolachi, C.; Pohoata, V.; Cimpoesu, N.; Bulai, G.; Iacomi, F.; Gurlui, S. Alpha keratin amino acids BEHVIOR under high FLUENCE laser interaction. Medical applications. *Appl. Surf. Sci.* **2019**, *488*, 418–426. [CrossRef]
15. Cocean, A.; Cocean, I.; Gurlui, S.; Iacomi, F. Study of the pulsed laser deposition phenomena by means of Comsol Multiphysics. *Univ. Politeh. Buchar. Sci. Bull. Ser. A Appl. Math. Phys.* **2017**, *79*, 263–274.
16. Pretsch, E.; Buhlmann, P.; Badertscher, M. Structure Determination of Organic Compounds: Tables of spectral data. *Struct. Determ. Org. Compd.* **2009**, *10*, 978–1007. [CrossRef]
17. Singh, J.; Kaur, N.; Kaur, P.; Kaur, S.; Kaur, J.; Kukkar, P.; Kumar, V.; Kukkar, D.; Rawat, M. Piper betle leaves mediated synthesis of biogenic SnO_2 nanoparticles for photocatalytic degradation of reactive yellow 186 dye under direct sunlight. *Environ. Nanotechnol. Monit. Manag.* **2018**, *10*, 331–338. [CrossRef]
18. Blanchard, M.; Méheut, M.; Delon, L.; Poirier, M.; Micoud, P.; Le Roux, C.; Martin, F. Infrared spectroscopic study of the synthetic Mg–Ni talc series. *Phys. Chem. Miner.* **2018**, *45*, 843–854. [CrossRef]
19. Ozkan, S.A.; Ahmed, R. (Eds.) *Novel Developments in Pharmaceutical and Biomedical Analysis*; Bentham Science Publishers: Sharjah, United Arab Emirates, 2018.
20. Van Steen, E.; Callanan, L.H. Part 3, Studies in Surface Science and Catalyzes. In *Recent Advances in the Science and Technology of Zeolites and Related Materials, Proceedings of the 14th International Zeolite Conference, Cape Town, South Africa, 25–30 April 2004*; Elsevier: Amsterdam, The Netherlands, 2004; Volume 154.
21. Hamdan, H.; Endud, S.; He, H.; Muhid, M.N.M.; Klinowski, J. Alumination of the purely siliceous mesoporous molecular sieve MCM-41 and its hydrothermal conversion into zeolite Na-A. *J. Chem. Soc. Faraday Trans.* **1998**, *24*, 3527–3780.
22. Zakaria, M.A.; Menazea, A.; Mostafa, A.M.; Al-Ashkar, E.A. Ultra-thin silver nanoparticles film prepared via pulsed laser deposition: Synthesis, characterization, and its catalytic activity on reduction of 4-nitrophenol. *Surf. Interfaces* **2020**, *19*, 100438. [CrossRef]

23. El Sayed, S.; De La Torre, C.; Santos-Figueroa, L.E.; Pérez-Payá, E.; Martínez-Máñez, R.; Sancenón, F.; Costero, A.M.; Parra, M.; Gil, S. A new fluorescent "turn-on" chemodosimeter for the detection of hydrogen sulfide in water and living cells. *RSC Adv.* **2013**, *3*, 25690. [CrossRef]
24. Boom, A.F.J.V.D.; Pujari, S.P.; Bannani, F.; Driss, H.; Zuilhof, H. Fast room-temperature functionalization of silicon nanoparticles using alkyl silanols. *Faraday Discuss.* **2020**, *222*, 82–94. [CrossRef] [PubMed]
25. Tsuboi, Y.; Kimoto, N.; Kabeshita, M.; Itaya, A. Pulsed laser deposition of collagen and keratin. *J. Photochem. Photobiol. A Chem.* **2001**, *145*, 209–214. [CrossRef]
26. Donaldson, L.; Radotic, K.; Kalauzi, A.; Djikanovic, D.; Jeremic, M. Quantification of compression wood severity in tracheids of Pinus radiate D. Don using confocal fluorescence imaging and spectral deconvolution. *J. Struct. Biol.* **2010**, *169*, 106–115. [CrossRef]
27. Radotić, K.; Kalauzi, A.; Djikanović, D.; Jeremić, M.; Leblanc, R.M.; Cerovic, Z.G. Component analysis of the fluorescence spectra of a lignin model compound. *J. Photochem. Photobiol. B Biol.* **2006**, *83*, 1–10. [CrossRef]

Article

Effect of Alkyl Structure (Straight Chain/Branched Chain/Unsaturation) of C18 Fatty Acid Sodium Soap on Skin Barrier Function

Koji Kubota [1,2,3,*], Akie Kakishita [2], Mana Okasaka [3,4], Yuka Tokunaga [3] and Sadaki Takata [3,4]

[1] Department of Pharmacy, Faculty of Pharmacy, Iryo Sosei University, Iwaki-City, Fukushima 971-8550, Japan
[2] Department of Pharmacy, Faculty of Pharmacy, Yasuda Women's University, Hiroshima-City, Hiroshima 731-0153, Japan; 11141108@g.yasuda-u.ac.jp
[3] Department of Fashion and Beauty Sciences, Faculty of Liberal Arts, Osaka Shoin Women's University, Higashiosaka-City, Osaka 577-8550, Japan; okasaka.mana@osaka-shoin.ac.jp (M.O.); tokunaga.yuka@osaka-shoin.ac.jp (Y.T.); takata.sadaki@osaka-shoin.ac.jp (S.T.)
[4] Division in Fashion and Beauty Studies, Graduate School of Human Sciences, Osaka Shoin Women's University, Higashiosaka-City, Osaka 577-8550, Japan
* Correspondence: koji.kubota@isu.ac.jp; Tel.: +81-24-629-5111

Received: 20 May 2020; Accepted: 17 June 2020; Published: 23 June 2020

Abstract: Anionic surfactants are commonly used as detergents and emulsifiers. However, these compounds are potent skin irritants. In this study, we evaluated the effect of the alkyl structure of anionic surfactants on the skin barrier function using the transmission index (TI) method. The TI method is used to measure the skin penetration rate of drugs. Sodium soaps of C18 fatty acids with different structures were evaluated. Sodium laurate was used as the control. In addition, microscopic observations of the skin tissue treated with different soaps and controls were performed to study the mechanism of skin permeation. Results showed that unsaturated fatty acid soaps exerted the most potent effect on the skin barrier function and saturated fatty acid soaps exerted the least effect; saturated branched fatty acid soap had an intermediate effect. This could be attributed to the differences in the melting points of different fatty acids. In addition, unlike lauric acid soap, C18 fatty acid soap did not cause morphological changes in the skin tissue. Thus, differences in the alkyl structure of fatty acids resulted in differences in the effect of fatty acid soaps on the skin barrier function. The mechanism was presumed to be an effect on intercellular lipids.

Keywords: surfactant; skin barrier function; skin permeability; alkyl structure

1. Introduction

Saponified fatty acids (soaps) are common anionic surfactants that are widely used as detergents and emulsifiers. Among fatty acid soaps, lauric acid soap is often used as a body cleanser, shampoo, dishwashing detergent, and laundry detergent, owing to its good foaming capacity and strong detergency. However, surfactants have been reported to affect the skin barrier function [1–3].

The skin barrier consists of keratin proteins and intercellular lipids. Keratin, a cytoskeletal protein, is composed of a group of intermediate filaments composed of keratinocytes, which are differentiated and enucleated epidermal cells. Intercellular lipids have a lamellar structure in which the lipid molecules, mainly composed of ceramide, fatty acids, and cholesterol esters, are regularly arranged. Stratum corneum cells and intercellular lipids form a structure known as the brick and mortar model, establishing a skin barrier function [4–6].

Previous studies on the effects of surfactants on the skin have reported a correlation with the alkyl chain length of surfactants [7–9]. Kurosaki et al. evaluated the relationship between the carbon number of fatty acid surfactants and irritancy to skin (using skin roughness and degreasing power in the patch test as evaluation scales), and revealed that C12 alkyl anionic surfactants caused the strongest skin irritation [10]. In addition, it has been reported that the skin barrier function is reduced due to the formation of a stratum corneum permeation path for a specific molecule, thereby increasing the skin permeability of the molecule. Using this phenomenon, the skin barrier function has been evaluated by measuring the invasiveness of substances or transepidermal water loss (TEWL) [6,11–13]. In vitro skin permeation assays, using the Franz diffusion cell system, are the most commonly used methods for assessing the skin absorption of drugs [14]. Okasaka et al. developed a transmission index (TI) method to evaluate the effect of anionic surfactants on the skin barrier function, using the skin permeation rate of methylparaben as an index [15].

Some previous studies have reported interesting findings on the effects of amphiphiles on the skin barrier function. Ionic liquids are amphiphilic substances similar to surfactants. Several interesting reviews of ionic liquids for enhancing drug skin permeation have been summarized [16,17]. Dobler et al. has reported the skin penetration effect by water-in-oil (W/O) or oil-in-water (O/W) emulsion, consisting of the hydrophobic ionic liquid or by the hydrophilic ionic liquid [18]. Some ionic liquids are formed using fatty acids as raw materials and are known to significantly enhance the skin permeability of various drugs. Medrx Co. Ltd., has commercialized an ionic liquid composed of fatty acids and amines with high biocompatibility as a skin permeation enhancer [19]. Kubota et al. investigated the mechanism by which fatty acid–aliphatic amine ionic liquids enhance the skin permeation of drugs [20]. Their results have shown that the skin penetration-promoting effect of drugs differs depending on the composition of the ionic liquid. While the ionic liquid composed of octanoic acid and triisopropanolamine markedly enhanced the skin permeability of hydrophilic model drugs, the ionic liquid derived from isostearic acid, a branched-chain C18 fatty acid, has an inhibitory effect on the skin permeability of hydrophobic drugs [20]. These results suggest that the structure of the alkyl chain of the amphiphile, including the surfactant, affects the skin barrier function. Hence, we investigated the effect of the alkyl structure of fatty acids on the skin barrier function using fatty acid soaps, which are common and simple amphiphiles. C18 fatty acids were used as the raw materials for the preparation of the fatty acid soaps. C18 fatty acids have many structural isomers with different branching and degree of unsaturation, and they are widely used as cosmetic or food ingredients or additives owing to their safety and high biocompatibility [21,22]. In this study, we used stearic acid as the straight chain fatty acid, isostearic acid as the branched-chain fatty acid, and oleic acid (18:1 (n-9)) and linoleic acid (18:2 (n-6)) as the unsaturated fatty acids. These fatty acids were neutralized with sodium hydroxide to prepare the saponified fatty acids. The effect of these structurally isomeric C18 fatty acid soaps on the skin barrier function was evaluated by the TI method described by Okasaka et al. [15] Skin permeability assay was performed on the back skin of hairless mice using the Franz diffusion cell system. Methylparaben was used as an indicator. In parallel, the skin tissue injury was evaluated by microscopic observation.

2. Materials and Methods

2.1. Preparation of the Soaps

Fatty acid soaps were prepared by saponification reaction between the fatty acid and sodium hydroxide. The C18 fatty acids were prepared based on the classification of the alkyl structure. Stearic acid was prepared as a saturated straight-chain fatty acid. The optical isomeric mixture of isostearic acid was prepared as a saturated branched-chain fatty acid. Two types of C18 fatty acids were prepared as unsaturated fatty acids. Oleic acid was prepared as a monounsaturated fatty acid, while linoleic acid was prepared as a di-unsaturated fatty acid. For comparison with these sodium soaps, sodium laurate was prepared from lauric acid. All reagents were of first or special grade and purchased from Wako Pure Chemical Industries (Tokyo, Japan). A 1.0 mol/L sodium hydroxide aqueous solution was prepared by dissolving solid sodium hydroxide (Sigma-Aldrich, St. Louis, MO, USA) in deionized water. A 5 mL solution of 1 mol/L sodium hydroxide (5 mmol NaOH) was stirred on a hot plate stirrer (AS ONE Co., Osaka, Japan) maintained at 70 °C. Then, 5 mmol fatty acid was added to it and the mixture was stirred until it was completely homogenized. The resultant saponified fatty acid was adjusted to a 1% concentration by adding deionized water.

2.2. Skin Permeation Experiments and Evaluation of the Effect of Fatty Acid Soap on the Skin Barrier Function

The effect of different fatty acid soaps on the skin barrier function was assessed by the TI method, based on a previously described method [15].

The skin permeability assay was performed using the Franz diffusion cell system (vertical reservoir, 8 mL; permeation area, 1 cm^2) (PermGear, Hellertown, PA, USA). The reservoir chamber was filled with saline and warmed up to 37 °C. Laboskin® (Hoshino Laboratory Animals Inc., Ibaraki, Japan; certificated by the Japanese Society for Laboratory Animal Resources); the back skin from a 7-week-old male Hos:HR-1 hairless mouse, was divided into four parts and used as the transmission skin sample. There were no statistically significant differences in the head and tail sides as well as the right and left sides of the midline. A 1 mL 1% fatty acid soap solution or control sample (deionized water) was applied to the external surface of the skin for 1 h. After treatment, the test or control sample was removed and the skin was thoroughly washed five times each with 1 mL of deionized water followed by 2 mL of deionized water.

Methylparaben was purchased from Ueno Fine Chemical Industry Ltd. (Osaka, Japan). One hundred milligrams of methylparaben was accurately weighed using an electronic analytical scale AUX120 (Shimadzu, Kyoto, Japan) and dissolved in deionized water. To ensure complete the dissolution of methylparaben, the solution was stirred on a vortex mixer, followed by ultrasonication for 15 min. It was then diluted with 100 mL of deionized water to obtain a 0.100% (w/v) solution. After skin stimulation with soap, 1 mL of the 0.100% methylparaben solution was applied to the skin after washing twice. The donor cup was sealed with parafilm to prevent the evaporation of the methylparaben solution. The reservoir solution was stirred at approximately 500 rpm, using either 6 or 12 stations multi magnetic stirrer (AS ONE). The reservoir solution in the receptor cup was mixed 10 times with a micropipette. After 1 h, 200 µL of the sample was withdrawn and an equivalent volume of saline was replaced. The process was repeated after every hour for 6 h.

The concentration of methylparaben in the collected reservoir solution was measured by High Performance Liquid Chromatography (HPLC) method using an LC-20AB HPLC system (Shimadzu) with a Capcell Pak® 5 µm C18 MG-II column (internal diameter: 2.0 mm × 150 mm) (Shiseido, Tokyo, Japan). An accurately diluted sample of 0.002% methylparaben was used as the standard. The column oven temperature was set at 40 °C. The HPLC mobile phase, consisting of (A) 15 mM phosphoric acid (Nakarai tesque, Kyoto, Japan) and (B) methanol (for HPLC, Wako), was adjusted (A:B = 55:45), and the flow rate was maintained at 0.2 mL/min. Five microliters of the reservoir solution was injected using an autosampler SIL-20AC (Shimadzu). The peak area of the UV absorbance was measured at 256 nm and calibrated using the LCsolution 1.23 SP1 software (Shimadzu). The reference sample (0.001% methylparaben) was prepared by diluting the sample from the permeability test and was analyzed before and after the analysis of the test sample, to confirm the stability of methylparaben during the analysis. No interferences were observed from the blank skin extracts during the HPLC analysis of the collected reservoir solutions. Each experiment was repeated at least 4 times and statistical analysis was carried out.

The TI value was calculated as follows. The $Flux_{ss}$ was determined from the quantity of permeate (Q) with respect to time (t), when the skin permeability indicated a linear correlation with time (Equation (1)). TI was determined as the relative value of the $Flux_{ss}$ of control and the $Flux_{ss}$ of the sample (Equation (2)). The $Flux_{ss}$ value was calculated using the LINEST function on the time-concentration scatter plot in Microsoft Excel and statistical analysis using Welch's t-test and Pearson's chi-square test were carried out:

$$Flux_{ss} = \frac{dQ}{dt} \quad (1)$$

$$\text{Transmission Index} = \frac{Flux_{ss}(surfactant)}{Flux_{ss}(control)} \qquad (2)$$

2.3. Assessment of Skin Tissue Injury

The skin tissue injury caused by the fatty acid soap was assessed by microscopic observation, according to the method described in our previous study [20]. The skin samples were treated with surfactant or control for 1 h before the assessment. After rough sectioning using surgical scissors, the skin samples were washed 3 times in 20 mL of Dulbecco's recipe of phosphate buffered saline without Mg^{2+} and Ca^{2+} (PBS(−)) (10 mM, pH 7.4) for 5 min each. The tissue samples were fixed in 4% paraformaldehyde/0.1 M phosphate buffer (pH 7.4), stored overnight, and immersed in 20% sucrose/0.1 M phosphate buffer (pH 7.4) at 4 °C. The tissue blocks were embedded in an optical cutting temperature compound (Tissue-Tek, Sakura Finetek; Torrance, CA, USA) and snap-frozen in powdered dry ice. The frozen embedded tissue samples were cut to approximately 10 μm slices and air-dried. Then, the tissue slices were deposited on glass slides, moistened with water, and stained with Mayer's hematoxylin (Wako Pure Chemical Industries) and eosin Y (Waldeck, Münster, Germany). After dehydration in graded anhydrous ethanol and Clear Plus (Falma Co.; Tokyo, Japan), a cover glass was used to seal the slices on the glass slides. Microscopic images were acquired using a BX51 bright field/fluorescence biological microscope (Olympus, Tokyo, Japan). The brightness and contrast of the images were automatically optimized using the ImageJ software.

3. Results

3.1. Changes during Preparation of the Aqueous Soap Solutions

The fatty acids that were liquid at room temperature (mixtures of optical isomers of isostearic acid, oleic acid, and linoleic acid) immediately changed to a white gel-like saponified product upon the addition of 1 mol/L sodium hydroxide solution. This saponified product was dissolved by stirring to form a colorless, transparent solution with a specific viscosity and a peculiar soapy smell. While preparing stearic acid soap, the saponified product was dissolved until about half of the original amount of the powdered stearic acid, which was added to the aqueous sodium hydroxide solution. Thereafter, on further stirring, the saponified product turned into a colloidal solution and finally became a high-viscosity colloidal solution containing white suspended matter. Upon incubation at 65 °C for 30 min, the suspended matter dissolved and a low-viscosity, slightly turbid solution, was produced. Below room temperature (25 °C), the viscosity of the incubated solution increased and the solution became almost non-fluid after standing overnight. In contrast, powdered lauric acid immediately dissolved in the sodium hydroxide solution and produced a low-viscosity, almost colorless, transparent solution.

3.2. Evaluation of the Effect of Sodium Fatty Acid Soaps on the Skin Barrier Function

The effects of the various fatty acid soaps on the skin barrier function were evaluated. Figure 1 shows the skin permeation profile of methylparaben in skin samples treated with the five fatty acid soaps and the control (deionized water). The change in the skin permeability of methylparaben showed a high correlation with the permeation time for all samples (R > 0.94), confirming that the skin permeation profile from 1 to 6 h was at a steady state.

Based on this observation, the effect of fatty acid soap on skin barrier function was evaluated using the TI method. The slopes of the time-concentration scatter plot for each fatty acid soap and control-treated sample are shown in Figure 2 as $Flux_{ss}$. The TI values of the five fatty acid soaps are shown in Table 1. The highest $Flux_{ss}$ value was observed for sodium laurate (6.08 ± 0.27). It was observed that the $Flux_{ss}$ of C18 fatty acid soap varied depending on the alkyl structure as follows: stearic acid (1.67 ± 0.05) < isostearic acid (2.38 ± 0.10) < oleic acid (4.44 ± 0.22) < linoleic acid (5.18 ± 0.1), saturated straight-chain fatty acid < saturated branched-chain fatty acids < monounsaturated fatty acids < diunsaturated fatty acids.

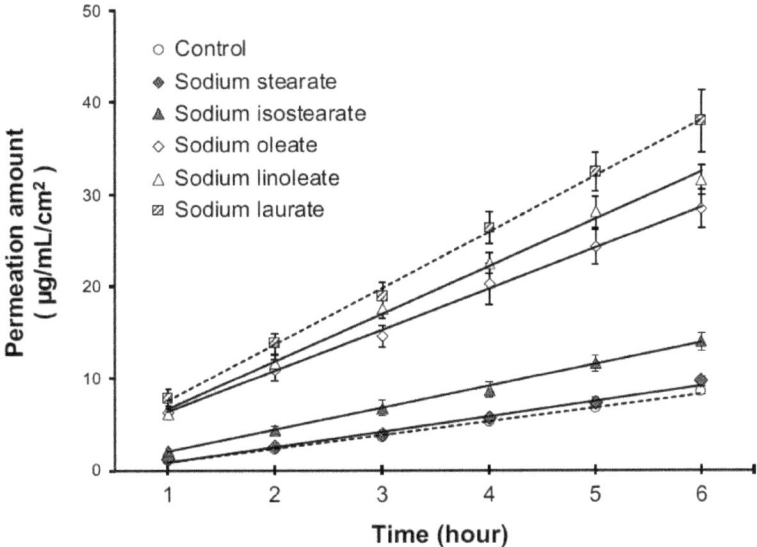

Figure 1. Time-dependent change in the permeation of methylparaben in skin stimulated with each surfactant and control (water). The skin permeation rate of methylparaben from 1 to 6 h was constant, indicating that the skin permeation was in a steady state. Each point value represents mean ± standard error, n = 35 (control), n = 6–10 (sample).

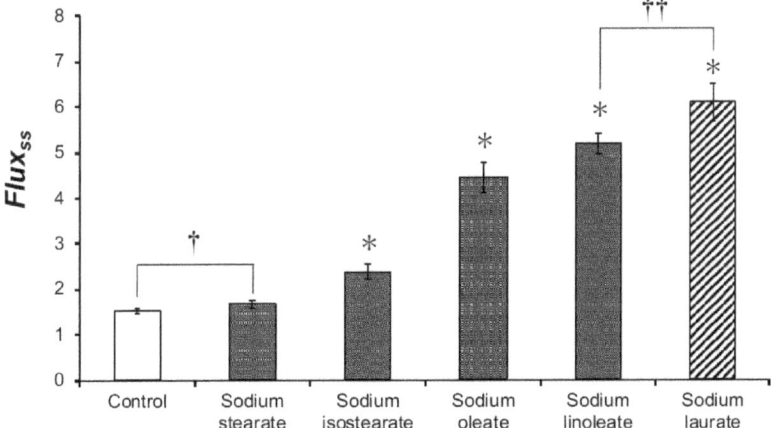

Figure 2. The Fluxss of the skin samples stimulated by each surfactant and control (water). Each point value represents the mean ± standard error, n = 35 (control), n = 6–10 (sample). The white bar represents the Fluxss of the water-treated sample (control). The shaded bar represents the Fluxss of the sodium laurate-treated sample. The Fluxss of each structurally isomeric C18 fatty acid soap (sodium stearate, sodium isostearate, sodium oleate, and sodium linoleate)-treated sample is represented by gray bars and the name of each fatty acid soap is described below the horizontal axis. The measured value, represented by symbol (*), shows the statistically significant difference against the control; $p < 0.05$. The non-statistically significant difference observed against the control is represented by † and against sodium laurate is represented by ††; $p > 0.1$.

Table 1. Transmission index (TI) value of each surfactant.

Fatty Acid Soap	Transmission Index ± SE
Sodium stearate	1.10 ± 0.08
Sodium isostearate	1.56 ± 0.10
Sodium oleate	2.92 ± 0.11
Sodium linoleate	3.40 ± 0.07
Sodium laurate	3.99 ± 0.10

In our previous study, we observed that surfactants with a significant difference in $Flux_{ss}$, with respect to the control, had an effect on the skin barrier function [15]. Since the TI value of stearic acid soap was 1.10 ± 0.08, which was not significantly different from the control $Flux_{ss}$, we concluded that it did not affect the skin barrier function. The TI value of the isostearic acid soap was 1.56 ± 0.10, which was significantly different from the control $Flux_{ss}$ ($p < 0.05$). The TI values of the two unsaturated C18 fatty acid soaps, prepared using oleic acid and linoleic acid as raw materials, were 2.92 ± 0.07 and 3.40 ± 0.07, respectively, and were significantly different from the control $Flux_{ss}$. The TI value of the lauric acid soap was extremely high at 3.99 ± 0.10. Furthermore, the $Flux_{ss}$ values of the oleic acid and linoleic acid soaps were not significantly different from those of lauric acid soap ($p > 0.1$). There was no significant difference in the TI values of oleic acid soap and linoleic acid soap ($p > 0.1$). However, linoleic acid soap with a higher unsaturation showed a higher TI value.

3.3. Assessment of Skin Tissue Injury

The microscopic images, depicting the degree of skin injury caused by various fatty acid soaps, are presented in Figure 3. The microscopic image of the normal skin tissue, without any treatment, is shown in Figure 3a. In normal tissue, the outermost keratin of the stratum corneum (stained with eosin) and healthy nuclei in the epidermal cell layer (stained with hematoxylin) were observed. Cells in the dermis layer were maintained in normal shape, as observed by eosin-stained collagen, and contained normal nuclei. The skin tissue section obtained after water treatment was similar to the untreated sample (Figure 3b). Results of the microscopic examination of the skin samples treated with each C18 fatty acid soap are shown in Figure 3c–g. The tissue morphologies of the skin treated with sodium stearate, sodium isostearate, sodium oleate, and sodium linoleate were similar to those of the untreated skin. However, the tissue sample treated with sodium laurate showed significant damage (Figure 3g). A part of the stratum corneum was separated from the epidermis layer and a gap was formed between the epidermis and the dermis. Chromatin was not observed in the epidermal cells. Chromatin loss was also found deep in the dermis, suggesting that the cell damage extended deep into the skin tissue.

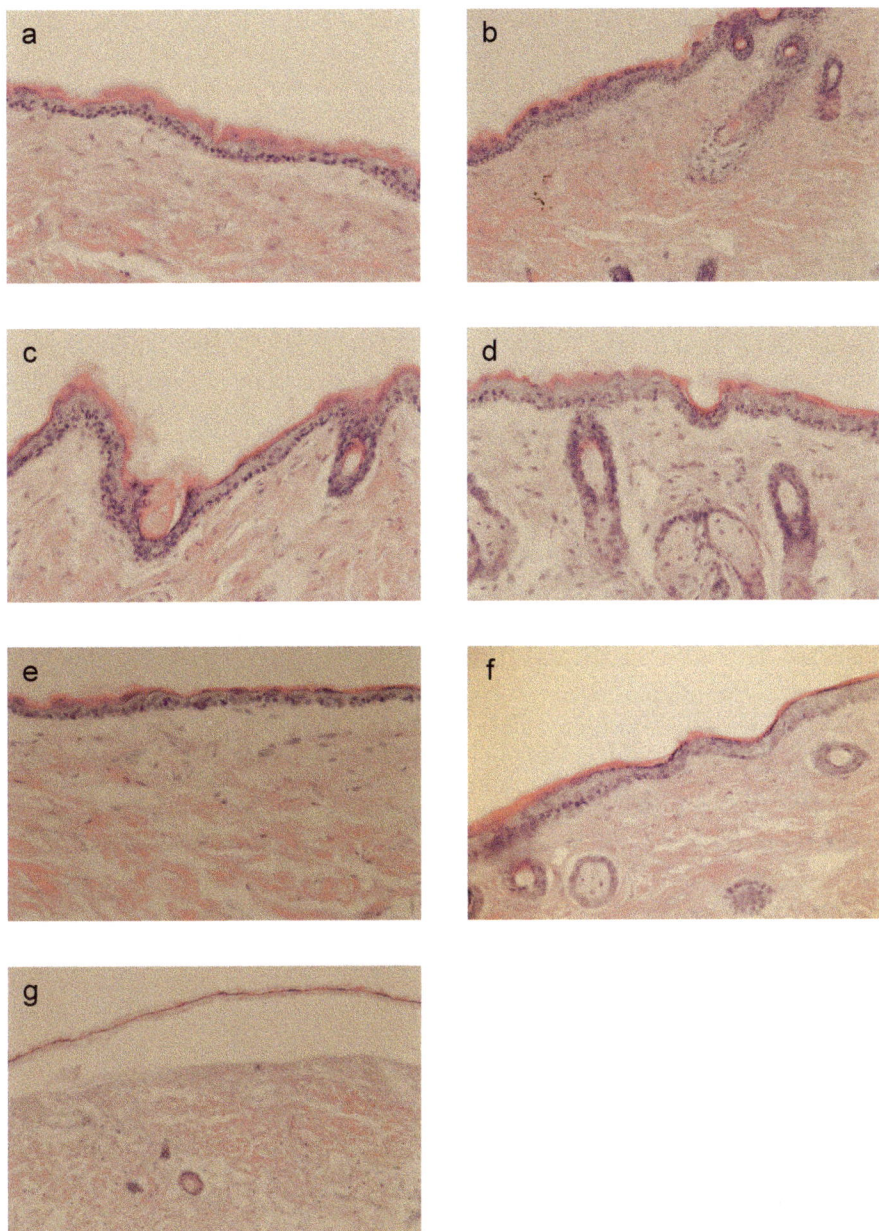

Figure 3. Micrograph of the skin tissues stimulated with 1% fatty acid soap after the hematoxylin–eosin (HE) staining. (**a**) represents the intact skin tissue, magnified with a 20× objective. Sample treated with water for 1 h is represented in (**b**). The skin tissue samples treated for 1 h with 1% (**c**) sodium stearate, (**d**) sodium isostearate, (**e**) sodium oleate, (**f**) sodium linoleate, and (**g**) sodium laurate. (**b**–**f**) are magnified with a 20× objective, and (**g**) with a 10× objective.

4. Discussion

In this study, the effect of the structural differences in the various isomers belonging to the hydrophobic group of anionic surfactants on the skin barrier function was evaluated using the TI method. This method was developed for the purpose of the easy analysis of the effect of surfactants on the skin barrier function. It is an effective method for the comparative study of surfactants, provided the drug has a steady state in skin permeation (Figure 1).

The TI value of the sodium laurate soap was approximately 4 (Table 1), which was different from the value obtained in our previous study (2.53) [15]. Of note, the sodium laurate soap assessed in the previous study was a cosmetic raw material-grade industrial product. On the other hand, the sodium laurate soap in this study was prepared using reagent-grade lauric acid and sodium hydroxide. Thus, it was assumed that the differences in purity and neutralization rates of the fatty acids led to the different TI values. From a qualitative point of view, previous studies have shown that sodium laurate soap has a significantly higher TI value than other anionic surfactants, indicating that the effect of sodium laurate soap on the skin barrier function is strong. Previous studies have reported that C12 fatty acid surfactants cause strong skin irritation [10]. Based on the fact that the surfactants with strong skin irritation have high TI values, it was suggested that the TI value was related to skin irritation.

The TI values are expressed as the ratio of $Flux_{ss}$ of surfactant-treated skin to control-treated skin. The standard error (SE) was calculated from the unbiased variance of the combination of the $Flux_{ss}$ ($n = 35$) of control and $Flux_{ss}$ ($n = 6$–10) of the test sample.

The significant differences in the TI values between the C18 fatty acid soaps suggested that the effect of sodium fatty acid soaps on the skin barrier function was affected not only by the carbon number of alkyl structures, but also by their alkyl structures. When the fatty acids were arranged in increasing order of their respective TI values (saturated straight chain < saturated branched chain < monounsaturated ~< di-unsaturated), the sequence corresponded to the respective melting points of the fatty acids. The melting points of the saturated straight-chain stearic acid, saturated branched-chain isostearic acid, monounsaturated oleic acid, and di-unsaturated linoleic acid are 70, 16, ≤15, and −5 °C, respectively. There is a relationship between the melting point and solubility of a substance, with substances with higher melting points generally being less soluble [23]. In soaps, generally, the Krafft point is proportionally affected by the melting point. The Krafft point increases as the number of carbon atoms in the alkyl chain increases. At the same time, if the hydrophilic groups are the same, the solubility in the water decreases as the carbon number increases. When an alkyl structure includes unsaturated groups, the Krafft point decreases and the water solubility increases. In addition, the solubility tends to increase as the hydrophilic group approaches the center of the alkyl chain [24]. The difference in the solubility of the sodium soap was also confirmed by the change in the dissolution state during the soap preparation as well as the properties of the prepared soap. Based on these facts, it was assumed that the melting point, Krafft point, and the solubility of the raw fatty acid were the main factors affecting the skin barrier function. The relationship between the carbon number of surfactants and skin irritation is well known [25,26]. It can be said that their report is a mention to one aspect of the relationship between the melting point of fatty acids and the properties of surfactants on the skin.

The temperature conditions of the experiment are important because the solubility is affected by temperature. The TI value was calculated based on the skin permeation speed of methylparaben at 37 °C. The equilibrium at this temperature depends on the solubility of the fatty acid sodium soap. We predicted that the higher the solubility of the surfactant, the stronger the action on the keratinocyte intercellular lipid, and as a result, the skin permeability of the surfactant would be enhanced. In addition, because fatty acid soap is a weak acid and strong alkalin salt, the equilibrium state depends on the pH. Furthermore, the critical micelle concentration is affected by the temperature and the pH of the aqueous solution. Therefore, it is necessary to discuss the influence of the physicochemical properties of the soap molecules on the skin barrier function, based on the measurements of critical micelle concentration or the pH dependence of the solubility.

In this study, we also examined the effect of anionic surfactant on the skin barrier function from a perspective other than drug skin permeability. The microscopic observation of hematoxylin–eosin (HE) stained tissue sections provided information on cytotoxicity and the effects of the surfactant on the connective tissue, based on chromatin staining and the condition of the connection between the tissues and cells. The absence of chromatin in the epidermal layer of the sodium laurate soap-treated skin (Figure 3g) indicated that the chromosome was damaged and the lipid bilayer of the cell membrane was probably damaged. This observation was similar to that seen upon the action of sodium lauryl sulfate (SDS), which exerts cytolytic effects due to its strong lipid solubility and protein denaturing activity [7]. This phenomenon associated with sodium laurate soap, also known as 'skin irritation' [10,27,28], results in the exfoliation of the stratum corneum, the lysis of epidermal cells, the disruption of the association between epidermal and dermal layers, and the cytotoxicity to cells in the dermal layers. The high TI value observed for the skin treated with sodium laurate soap could be attributed to the loss of the barrier function due to the physical damage of the skin tissue. However, although the unsaturated C18 fatty acid sodium soap had a TI value similar to that of the sodium laurate soap, no morphological damage to the skin tissue was observed (Figure 3e,f). This suggested that the mechanism of the effect of unsaturated C18 fatty acid sodium soap on skin barrier function was different from that of sodium laurate soap.

Since microscopic examination does not provide information regarding the changes in intercellular lipids, the effect of unsaturated C18 fatty acid sodium soap on skin barrier function may be due to changes in intercellular lipids. At present, we do not have information to link the TI values with the effects of two important factors on the skin barrier function, namely stratum corneum protein and the keratinocyte intercellular lipid. The inferences from the inconsistency between the results of microscopic observations and the trends in TI values of each surfactant need to be separately experimentally verified. That is our task in the future.

According to the structural chemistry considerations, it is considered that the effect of surfactants on the skin barrier function represented by the TI value is affected by the melting point of the fatty acids in the soap molecule and the solubility of the soap affected by them. Some previous studies have discussed the effects of ion pair properties of anionic surfactants on the skin [9,29–31]. However, in this study, the effects of counter ion and the neutralization rate on the TI value were not examined. They are also important factors affecting soap solubility. Soap solubility and its effects on skin barrier function need to be further investigated. Generally, potassium soap has a higher solubility than sodium soap. The neutralization rate also affects the pH of the aqueous soap solution. This study only provides information on the effects of the fatty acid branching structure and the degree of unsaturation in fully neutralized sodium soap. Moreover, the information available from microscopy is very limited. For future studies, an investigation of the relationship between the ion pair properties of the surfactants and the skin barrier function is warranted. This can contribute to the development of the manufacturing method such as the solubility and the pH adjustment of the surfactant which is safe for the skin barrier function and the suggestion of the direction for uses.

5. Conclusions

The investigation of the effect of the C18 alkyl anionic surfactants on the skin barrier function showed that the alkyl structures of anionic surfactant was a strong efficient factor for skin barrier function. Based on structural chemistry considerations, we noticed a correlation between the Kraft points and the C18 fatty acid soaps with different structures. On the other hand, interestingly, the difference among the strength of the effect of each C18 fatty acid soap on the skin barrier function does not correlate with the skin tissue injury. From this, it is possible that the effect of fatty acid soap on the skin barrier function can be controlled by the alkyl structure rather than the alkyl carbon number. This will be important information on the effect of the properties of matter of soap during hand washing and sterilization, or on its properties as a percutaneous absorption enhancer. However, since the experiment used skin extracted from mice, it is necessary to pay attention to the difference from the actual application to human skin. From the measurement of the TI value and microscopic observation, it is impossible to infer the mechanism of the influence of the difference in the alkyl structure of the surfactant on the skin barrier function. We strongly suspect the effects on interceluller lipids, but these have not yet been confirmed. Studies on the effects of keratinocyte lipids are our future work.

Author Contributions: K.K. and M.O. developed the Transmission Index method. And K.K. executed data analysis of skin permeability and microscopy experiments and wrote the entire manuscript and correspond for this article. A.K. conducted skin permeation experiments on the soaps mainly and Y.T. executed a part of experiments. S.T. is the general leader of the research project on the effect of surfactant on skin barrier function. All authors have read and agreed to the published version of the manuscript.

Funding: A part of this work was supported by JSPS KAKENHI Grant Number JP19K14014.

Conflicts of Interest: The authors declare no conflict of interest.

References

1. Ridout, G.; Hinz, R.S.; Hostynek, J.J.; Reddy, A.K.; Wiersema, R.J.; Hodson, C.D.; Lorence, C.R.; Guy, R.H. The effects of zwitterionic surfactants on skin barrier function. *Fund. Appl. Toxicol.* **1991**, *16*, 41–50. [CrossRef]
2. Lemery, E.; Briançon, S.; Chevalier, Y.; Oddos, T.; Gohier, A.; Boyron, O.; Bolzinger, M.-A. Surfactants have multi-fold effects on skin barrier function. *Eur. J. Dermatol.* **2015**, *25*, 424–435. [CrossRef] [PubMed]
3. Yanase, K.; Hatta, I. Disruption of human stratum corneum lipid structure by sodium dodecyl sulphate. *Int. J. Cosmet. Sci.* **2018**, *40*, 44–49. [CrossRef]
4. Elias, P.M.; Goerke, J.; Friend, D.S. Mammalian epidermal barrier layer lipids: Composition and influence on structure. *J. Investig. Dermatol.* **1977**, *69*, 535–546. [CrossRef]
5. Norlén, L. Skin barrier structure and function: The single gel phase model. *J. Investig. Dermatol.* **2011**, *117*, 830–836. [CrossRef]
6. Ruela, A.L.M.; Perissinato, A.G.; Lino, M.E.S.; Mudrik, P.S.; Pereira, G.R. Evaluation of skin absorption of drugs from topical and transdermal formulations. *Braz. J. Pharm. Sci.* **2016**, *52*, 527–544. [CrossRef]
7. Moore, P.M.; Puvvada, S.; Blankshtein, D. Challenging the surfactant monomer skin penetration model: Penetration of sodium dodecyl sulfate micelles into the epidermis. *J. Cosmet. Sci.* **2003**, *54*, 29–46.
8. Rhein, L.D.; Schlossman, M.; O'Lenick, A.; Somasundaran, P. *Surfactants in Personal Care Products and Decorative Cosmetics*, 3rd ed.; CRC Press, Taylor & Francis Group: Boca Raton, FL, USA, 2006.
9. Morris, S.A.V.; Thompson, R.T.; Glenn, R.W.; Ananthapadmanabhan, K.P.; Kasting, G.B. Mechanisms of anionic surfactant penetration into human skin: Investigating monomer, micelle and submicellar aggregate penetration theories. *Int. J. Cosmet. Sci.* **2019**, *41*, 55–66. [CrossRef]
10. Kurosaki, T.; Imokawa, G.; Ishida, A. The development of monoalkyl phosphate as a low skin irritating anionic surfactant. *J. Jpn. Oil Chem. Soc.* **1987**, *36*, 629–637. [CrossRef]
11. Roskos, K.V.; Guy, R.H. Assessment of skin barrier function using transepidermal water loss: Effect of age. *Pharm. Res.* **1989**, *6*, 949–953. [CrossRef] [PubMed]
12. Lee, S.H.; Jeong, S.K.; Ahn, S.K. An update of the defensive barrier function of skin. *Yonsei Med. J.* **2006**, *47*, 293–306. [CrossRef] [PubMed]
13. Gupta, J.; Grube, E.; Ericksen, M.B.; Stevenson, M.D.; Lucky, A.W.; Sheth, A.P.; Assa'Ad, A.H.; Hershey, G.K.K. Intrinsically defective skin barrier function in children with atopic dermatitis correlates with disease severity. *J. Allergy Clin. Immunol.* **2008**, *121*, 725–730. [CrossRef] [PubMed]
14. OECD. OECD guidelines for the testing of chemicals. Section 4: Health effects. In *Test. No. 428: Skin Absorption: In Vitro Method*; OECD Publishing: Paris, France, 2004.
15. Okasaka, M.; Kubota, K.; Yamasaki, E.; Yang, J.; Takata, S. Evaluation of anionic surfactants effects on the skin barrier function based on skin permeability. *Pharm. Dev. Technol.* **2019**, *24*, 99–104. [CrossRef]
16. Caparica, R.; Júlio, A.; Mota, J.P.; Almeida, C.R.T.S. Applicability of ionic liquids in topical drug delivery systems: A mini review. *J. Pharm. Clin. Res.* **2018**, *4*, 555649–555655. [CrossRef]

17. Sidat, Z.; Marimuthu, T.; Kumar, P.; Du Toit, L.C.; Kondiah, P.P.D.; Choonara, Y.E.; Pillay, V. Ionic liquids as potential and synergistic permeation enhancers for transdermal drug delivery. *Pharmaceutics* **2019**, *11*, 96. [CrossRef] [PubMed]
18. Dobler, D.; Schmidts, T.; Klingenhöfer, I.; Runkel, F. Ionic liquids as ingredients in topical drug delivery systems. *Int. J. Pharm.* **2013**, *441*, 620–627. [CrossRef]
19. Medrx Co., Ltd. External Preparation Composition Comprising Fatty Acid-Based Ionic Liquid as Active Ingredient. US 2014/0066471 A1, 6 March 2014.
20. Kubota, K.; Shibata, A.; Yamaguchi, T. The molecular assembly of the ionic liquid/aliphatic carboxylic acid/aliphatic amine as effective and safety transdermal permeation enhancers. *Eur. J. Pharm. Sci.* **2016**, *86*, 75–83. [CrossRef]
21. Bialek, A.; Bialek, M.; Jelinska, M.; Tokaz, A. Fatty acid profile of new promising unconventional plant oils for cosmetic use. *Int. J. Cosmet. Sci.* **2015**, *38*, 382–388. [CrossRef]
22. Michalak, M.; Kiełtyka-Dadasiewicz, A. Nut oils and their dietetic and cosmetic significance: A review. *J. Oleo Sci.* **2019**, *68*, 111–120.
23. Atkins, P.; Paula, J.D.; Keeler, J. *Atkins' Physical Chemistry*, 11th ed.; Oxford University Press: New York, NY, USA, 2017.
24. Nakama, Y. Surfactants. In *Cosmetic Science and Technology*; Elsevier: Amsterdam, The Nederland, 2017; pp. 231–244.
25. Potts, R.O.; Guy, R.H. A predictive algorithm for skin permeability: The effects of molecular size and hydrogen bond activity. *Pharm. Res.* **1995**, *12*, 1628–1633. [CrossRef]
26. Yamaguchi, F.; Watanabe, S.; Harada, F.; Miyake, M.; Yoshida, M.; Okano, T. In vitro analysis of the effect of alkyl-chain length of anionic surfactants on the skin by using a reconstructed human epidermal model. *J. Oleo Sci.* **2014**, *63*, 995–1004. [CrossRef] [PubMed]
27. Kanikkannan, N.; Singh, M. Skin permeation enhancement effect and skin irritation of saturated fatty alcohols. *Int. J. Pharm.* **2002**, *248*, 219–228. [CrossRef]
28. James-Smith, M.A.; Hellner, B.; Annunziato, N.; Mitaragotri, S. Effect of surfactant mixture on skin structure and barrier properties. *Ann. Biomed. Eng.* **2011**, *39*, 1215–1223. [CrossRef] [PubMed]
29. Benrraou, M.; Bales, B.L.; Zana, R. Effect of the nature of the counterion on the properties of anionic surfactants. 1. CMC, ionization degree at the CMC and aggregation number of micelles of sodium, cesium, tetramethylammonium, tetraethylammonium, tetrapropylammonium, and tetrabutylammonium dodecyl sulfates. *J. Phys. Chem. B* **2003**, *107*, 13432–13440.
30. Smith, G.N.; Brown, P.; James, C.; Kemp, R.; Khan, A.M.; Plivelic, T.S.; Rogers, S.E.; Eastoe, J. The effects of counterion exchange on charge stabilization for anionic surfactants in nonpolar solvents. *J. Colloid Interface Sci.* **2016**, *465*, 316–322. [CrossRef]
31. Yadav, S.K.; Kumar, S. Counterion-specific clouding in aqueous anionic surfactant: A case of Hofmeister-like series. *Colloid Polym. Sci.* **2017**, *295*, 869–876. [CrossRef]

© 2020 by the authors. Licensee MDPI, Basel, Switzerland. This article is an open access article distributed under the terms and conditions of the Creative Commons Attribution (CC BY) license (http://creativecommons.org/licenses/by/4.0/).

Article

Self-Expandable Retainer for Endoscopic Visualization in the External Auditory Canal: Proof of Concept in Human Cadavers

Yehree Kim [1,†], Jeon Min Kang [2,†], Ho-Young Song [3,4], Woo Seok Kang [1], Jung-Hoon Park [2,*,‡] and Jong Woo Chung [1,*,‡]

1   Department of Otorhinolaryngology-Head & Neck Surgery, Asan Medical Center, University of Ulsan College of Medicine, 88 Olympic-ro 43-gil, Songpa-gu, Seoul 05505, Korea; yehreek@hotmail.com (Y.K.); entkang7@gmail.com (W.S.K.)
2   Biomedical Engineering Research Center, Asan Institute for Life Scineces, Asan Medical Center, 88 Olympic-ro 43-gil, Songpa-gu, Seoul 05505, Korea; miny2208@naver.com
3   Department of Radiology, Asan Medical Center, University of Ulsan College of Medicine, 88 Olympic-ro 43-gil, Songpa-gu, Seoul 05505, Korea; hysong@amc.seoul.kr
4   Department of Radiology, UT Health Science Center at San Antonio, 7703 Floyd Curl Drive, San Antonio, TX 78229, USA
*   Correspondence: jhparkz@amc.seoul.kr (J.-H.P.); jwchung@amc.seoul.kr (J.W.C.)
†   These authors contributed equally to this work.
‡   These authors contributed equally to this work.

Received: 12 February 2020; Accepted: 4 March 2020; Published: 10 March 2020

Abstract: This study was conducted to investigate the efficacy of a self-expandable retainer (SER) for endoscopic visualization of the external auditory canal (EAC). Tympanomeatal flap (TMF) elevation was performed in six cadaveric heads. Two different types of SER were placed. The procedural feasibility was assessed by endoscopic images. Technical success rate, procedure time, endoscopy lens cleaning, and presence of mucosal injuries were analyzed. TMF elevation and SER placement were successful in all specimens and there were no procedure-related complications. The mean procedure time with the SERs was significantly shorter than without ($p < 0.001$). The mean number of times the endoscopy lens was cleaned during the procedure was significantly lower in the SER group ($p < 0.001$). In the SER group, endoscopy insertion into the EAC was easier without tissue contact with the lens during the TMF elevation compared with the non-SER group. There were no mucosal injuries. SER placement is effective for endoscopic visualization via the expanded and straightened EAC. A fully covered type of SER is preferable. The device can be useful for endoscopic ear surgery, reducing procedure time and reducing the need for endoscopy lens cleaning during the procedure.

Keywords: self-expandable metallic stent; ear canal; cadaver; ear speculum; endoscopy

1. Introduction

The past three decades have witnessed a growth in the use of endoscopy and it is used in several surgical disciplines including otologic surgery [1]. Endoscopic ear surgery (EES) has several advantages over conventional surgical microscopy. There is a wider field of view, improved resolution with high magnification, and the ability to "see around corners," allowing direct visualization of normally hidden recesses [2]. The external auditory canal (EAC) is typically used as the approach for surgery, obviating the need to create access pathways or retract soft tissues [3–5].

The EAC is about 2.5 cm in length and comprises a lateral cartilaginous portion and a medial bony portion [6,7]. There is a natural curvature to the external meatus which plays little role in endoscopy since the endoscope is positioned medial to the curvature [8]. However, the cartilaginous and bony portions need to be aligned to insert the endoscope. The cartilaginous portion of the EAC is often pushed posteriorly using the shaft of the endoscope. Certain techniques such as the Tarabichi's stitch can be used to straighten the EAC and increase the working space for the endoscope [9,10].

Because the space in the EAC and middle ear is limited, a clean surgical field is required to assure patient safety and benefit from the wide-angle vision offered by the endoscope [11]. The lens of the endoscope is easily smeared by the hair and earwax in the EAC and blood, however minimal, can obscure the view. The authors, therefore, devised a self-expandable retainer (SER) which can be placed in the EAC to aid surgery.

An ideal retainer would provide straight and clean entry to the middle ear without causing damage to the soft tissues of the EAC and, further, protect the tissues from trauma by the surgical instruments. The aim of this study was to investigate the efficacy and technical feasibility of SERs for endoscopic visualization in the EAC of cadaveric heads. SER placement was found to be effective for endoscopic visualization due to the expansion and straightening of the EAC.

2. Materials and Methods

2.1. Study Design

Twelve EACs of six adult human cadaveric heads were used in this study. None of them had a history of ear surgery, ear disease, or trauma. A tympanomeatal flap (TMF) was elevated in both ears of all the cadaveric specimens. One of the two available types of SER was placed in the left ear before the TMF elevation was performed. The bare type was placed in three cadaveric heads and the fully covered type of SER was placed in the remaining three. The right-sided ear had just the TMF elevation procedure without the placement of an SER.

2.2. Self-Expandable Retainers

Two different types of SERs were used (S&G Biotech, Yongin, Korea) (Figure 1): bare and fully covered types.

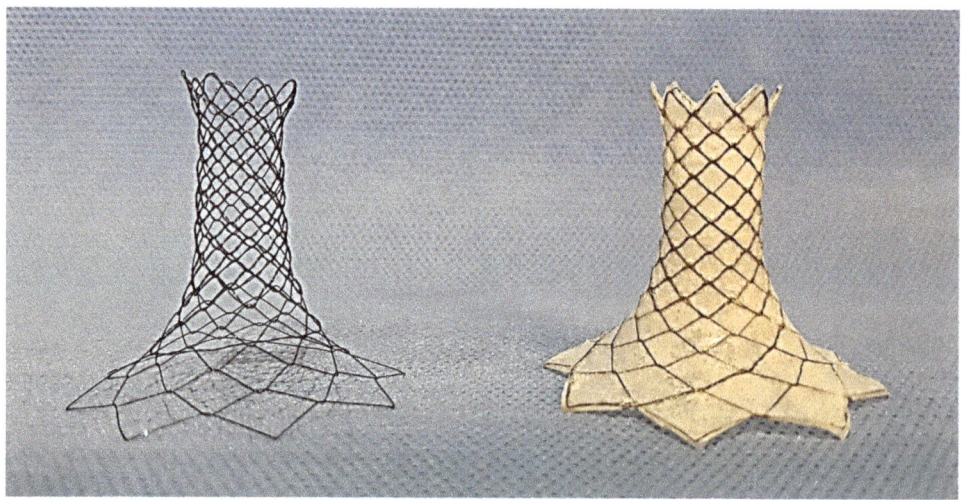

Figure 1. Photograph showing self-expandable retainers (SERs). Left: bare type of SER. Right: fully covered type of SER.

The SER comprises two parts: a straight self-expandable metallic stent (SEMS) part for the EAC and a flared part for the pinna. The straight SEMS part is knitted from a 0.127-mm nitinol wire into a tubular configuration. When fully expanded, the straight SEMS is 6 mm in diameter and 12 mm in length. The flared end, designed to be in direct contact with the pinna for endoscope insertion, is a round flared mesh made from a 0.127-mm nitinol wire. The end of the flared section is 40 mm in diameter and 23 mm in length and is spirally connected to the straight SEMS part. The total length of the SER is 35 mm. A fully covered variant of the SER is coated with silicone (MED-6640, Carpinteria, CA, USA) using the dipping method. The SERs were placed using forceps.

2.3. Tympanomeatal Flap Elevation and SER Placement

The cadaver head was turned to the side position. Using forceps, a collapsed SER was advanced through the pinna of the left EAC under endoscopic guidance (VISERA 4K UHD Rhinolaryngoscope; Olympus, Tokyo, Japan). The proximal end of the SER was located just lateral to the tympanic membrane. The TMF elevation using a canal knife was performed under endoscopic guidance with SER in place (left ear) or without SER (right ear). The SERs were immediately removed using forceps after the TMF elevation. Each specimen was carefully evaluated for any procedure-related complications using post-procedural endoscopic examination.

2.4. Study Definition and Data Analysis

The efficacy and feasibility of the use of SERs were assessed by the technical success rate, procedure time, number of times the endoscopy lens was cleaned, and presence of any mucosal injuries after SER removal, in the cadaveric specimens. Technical success was defined as successful TMF elevation under endoscopic guidance. The procedure time was measured from the insertion of the endoscope into the EAC (with or without the SER) to its removal after completion of the TMF elevation. The procedural feasibility and proximal end location of the SER were analyzed using endoscopic images.

2.5. Statistical Analysis

Data are expressed as the mean ± standard deviation (SD). The differences between the two groups were analyzed using the paired t test. The mean differences and 95% confidence intervals (CIs) were constructed, as appropriate. A p value of <0.05 was considered statistically significant. Statistical analyses were performed using SPSS software, version 23.0 (SPSS, IBM, Chicago, IL, USA).

3. Results

TMF elevation was technically successful in all specimens of the adult human cadaveric heads and there were no procedure-related complications. The placement of both types of SER in the left cadaveric ears was technically successful. The proximal ends of the SERs were located 2–3 mm from the tympanic membrane. The mean (±SD) procedure time for TMF elevation with SERs was significantly shorter than that for TMF elevation without SERs (175.6 ± 10.3 s vs. 312.2 ± 23.4 s; $p < 0.001$) (mean difference, 136.5 s; 95% CI, 111.7–161.3 s). The mean (±SD) number of endoscopy lens cleanings during the procedure was significantly less in the SER group compared with the non-SER group (0.83 ± 0.75 vs. 5.16 ± 1.17; $p < 0.001$) (mean difference, 4.33; 95% CI, 3.07–5.59) (Figure 2).

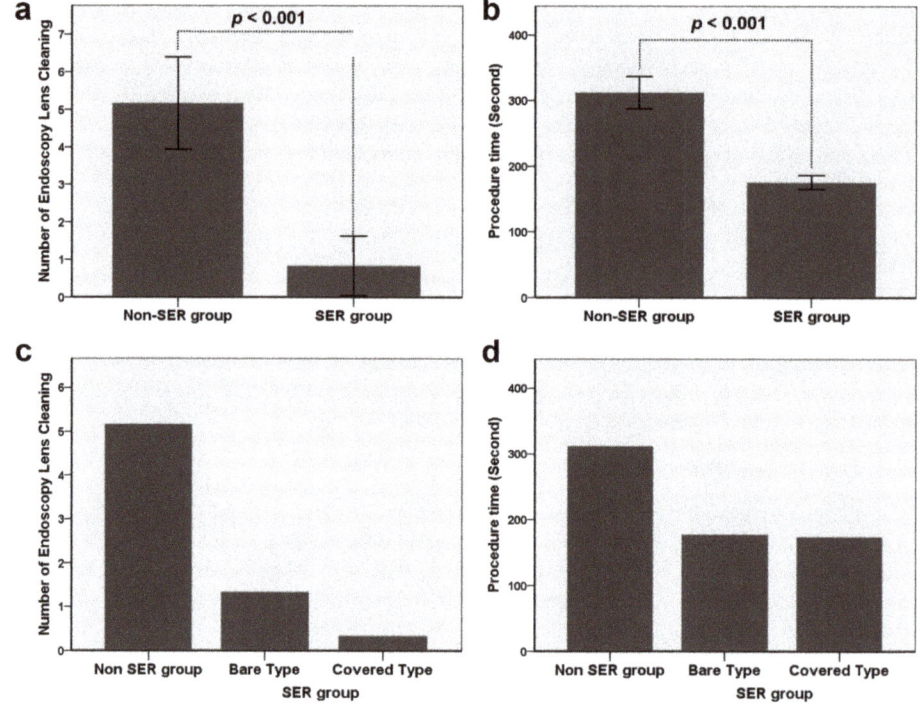

Note. SER: self-expandable retainer

Figure 2. The number of endoscopy lens cleaning interventions (**a**, **c**) and procedure time (**b**, **d**) in the non-SER and SER groups. Data are mean difference with a 95% confidence interval.

In the SER group, insertion of the endoscope into the EAC was relatively easy without any resistance or tissue contact by the lens during the TMF elevation compared with the non-SER group because of the expanded and straightened EAC (Figure 3).

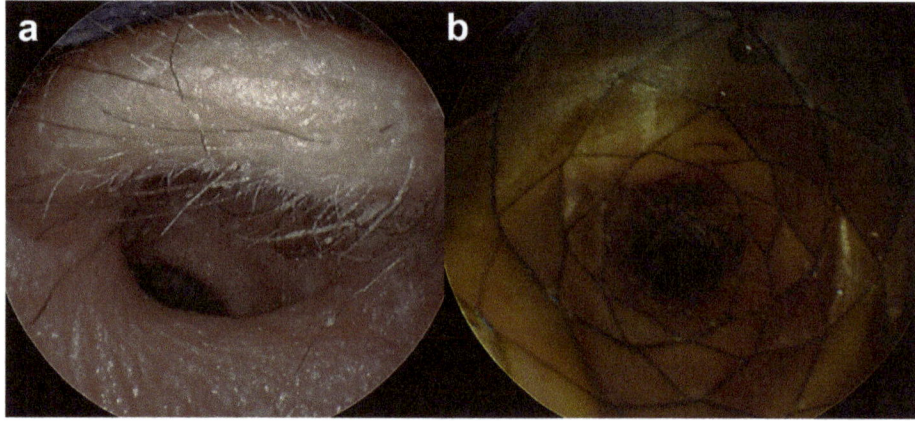

Figure 3. Endoscopic images showing the view at the pinna. (**a**) Endoscopic image showing hairs and irregular external auditory canal (EAC) anatomy. (**b**) Endoscopic image taken immediately after fully covered self-expandable retainer placement showing the expanded and straightened EAC.

However, with the bare type of SER, hairs and foreign bodies in the EAC were visible through the wire mesh of the straight part of the SEMS, (Figure 4).

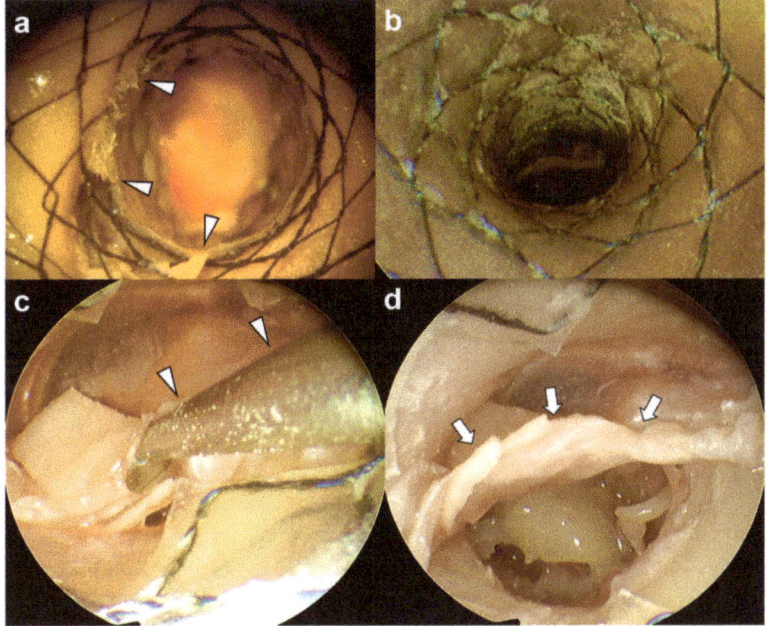

Figure 4. Endoscopic images obtained immediately after tympanomeatal flap (TMF) elevation in the self-expandable retainer (SER) group. (**a**) Endoscopic image showing the bare type of SER and hairs and earwax (arrowheads) in the external auditory canal visible through the wire mesh of the SER. (**b**) Endoscopic image showing the fully covered SER. (**c**) TMF elevation was performed using a canal knife (arrowheads). (**d**) A successfully elevated tympanic membrane (arrows).

The mean number of times the endoscopy lens was cleaned during the procedure was less in the fully covered SERs group (0.3 ± 0.56) compared with the bare SERs group (1.3 ± 0.57) without a statistical difference ($p = 0.101$). There were no mucosal injuries after the procedure in any specimen.

4. Discussion

The current study is a cadaveric proof of concept study of SER for better endoscopic visualization in EES. The newly developed SER has several advantages over the conventional ear speculum. First, it is safe and easy to place and remove with the use of forceps. After completing the whole procedure, endoscopic examination revealed no damage to the skin or soft tissue of the EAC.

Second, the use of an SER may allow a significant reduction in the procedure time during various EES due to less frequent endoscopic lens cleansing to remove hair and earwax. As EES is performed by a single surgeon, endoscope contamination can compromise the surgeon's vision and negatively affect the surgery [12]. In a survey asking otologists who routinely perform EES about the need for better instrumentation, more than 60% of respondents felt the need for improvements in order to 'keep the endoscope lens clean' [13].

The skin which lines the cartilaginous section of the canal is thicker, more mobile, and contains sebaceous and ceruminous glands and hair follicles [14] that can contaminate the lens while inserting the endoscope, requiring the surgeons to intermittently discontinue surgery to remove and clean the endoscope and reinsert it [15,16]. This can be time-consuming and taxes the surgeon's patience. Often the canal hair is trimmed before the start of the surgery to reduce fogging and smearing.

Endoscope contamination can also be due to bleeding. In traditional microscopic surgery, the non-dominant hand holds the suction instrument to maintain retraction and removal of blood from the operative field while the dominant hand holds the instruments [17]. Bleeding in EES is a major concern for the operating surgeon because the non-dominant hand also must hold the endoscope. Techniques utilized to achieve hemostasis include diathermy probes (monopolar and bipolar) [11], diluted epinephrine injection, epinephrine-soaked cotton balls that can also be used as adjuncts for soft tissue dissection [18], and preoperative injection of tranexamic acid [12] which is an anti-fibrinolytic used in a variety of surgical procedures [19–21].

Because our study used cadavers, we could not assess the advantage of the SER with regard to bleeding. The most common site of bleeding is in the posterior and superior portion of the EAC [11], where the main vascular supply comes from the anterior tympanic and deep auricular branches of the internal maxillary artery, the superficial temporal artery, and the posterior auricular artery [22]. We hypothesize that the SER might reduce bleeding by protecting the EAC skin from trauma from sharp surgical instruments and also by pressing against the vascular structures and reducing blood flow by the outward expanding force of the SER putting pressure on the soft tissues of the EAC. This idea needs further verification in preclinical and clinical studies.

Thermal injuries can be caused by the light source of endoscopes [23–28], therefore frequent removal of the endoscope from the field may be beneficial in preventing this [26]. Using an SER will reduce the frequency of endoscope removal from the surgical field; therefore other cooling methods should be implemented to ensure safety. These include the use of suction, turning off the light source at regular intervals, and the use of endoscope sheaths and saline irrigation [26,29].

SER not only allows for a straightened and slightly widened ear canal but also protects the soft tissue around the ear canal from the surgical instruments used in EES. While endoscopes facilitate a broader field of view, bone removal may still be necessary for better visualization and easier access. Osteotomes, curettes, drills, and piezoelectric systems are used for this [13]. When using drills, surgeons have to take care to protect the EAC soft tissue from possible injury by the drill shaft and an SER could be helpful with this.

There are some limitations to this study. First, although the variables of interest reached statistical significance, the sample size was too small to perform a robust statistical analysis. However, the differences in our findings between the SER and non-SER groups were indisputable. Second, we only performed endoscopic inspection to evaluate mucosal injuries after SER removal. Histopathological analysis needs to be used to verify possible mucosal injury. Third, in EES, one side could be easier to work with than the other, according to the surgeon. This matter of dexterity could also have affected the outcome as all SERs were inserted on the left ear in this cadaver study.

To our knowledge, we are the first to devise a tool for protection of the endoscope from contamination and the EAC from surgical trauma to assist in EES. A retainer can be helpful particularly in cases in which a long procedure time or bleeding is expected. Although additional studies are required to investigate the safety and optimal size of the SER, the results of our study support the basic concept of using an SER to enhance the space of the EAC for EES.

5. Conclusions

SER placement appears to be effective for endoscopic visualization by expanding and straightening the EAC in a human cadaveric head. A fully covered SER is preferable and the device has great potential for use in EES to reduce procedure time and reduce the frequency of endoscope lens cleaning during surgery.

Author Contributions: J.-H.P. and J.W.C. contributed to this work in study planning, cadaver experiment, data analysis, and manuscript preparation. Y.K. contributed to data analysis and manuscript preparation with critical comments; J.M.K., H.-Y.S., and W.S.K. contributed to the cadaver experiment. All authors have read and agreed to the published version of the manuscript.

Funding: This research was supported by a grant (2019IE7044) from the Asan Institute for Life Sciences, Asan Medical Center, Seoul, Korea.

Acknowledgments: The authors would like to thank Enago (http://www.enago.co.kr) for the English language review.

Conflicts of Interest: The authors declare no conflict of interest.

References

1. Kapadiya, M.; Tarabichi, M. An overview of endoscopic ear surgery in 2018. *Laryngoscope Investig. Otolaryngol.* **2019**, *4*, 365–373. [CrossRef] [PubMed]
2. Kiringoda, R.; Kozin, E.D.; Lee, D.J. Outcomes in Endoscopic Ear Surgery. *Otolaryngol. Clin N. Am.* **2016**, *49*, 1271–1290. [CrossRef] [PubMed]
3. Tarabichi, M. Endoscopic management of acquired cholesteatoma. *Am. J. Otol.* **1997**, *18*, 544–549. [PubMed]
4. Tarabichi, M. Endoscopic middle ear surgery. *Ann. Otol. Rhinol. Laryngol.* **1999**, *108*, 39–46. [CrossRef]
5. Mitchell, S.; Coulson, C. Endoscopic ear surgery: A hot topic? *J. Laryngol. Otol.* **2017**, *131*, 117–122. [CrossRef]
6. Anson, B.J.; Donaldson, J.A. *Surgical Anatomy of the Temporal Bone*; Saunders: Philadelphia, PA, USA, 1981; Volume 31.
7. Bojrab, D.I.; Bruderly, T.; Abdulrazzak, Y. Otitis externa. *Otolaryngol. Clin. N. Am* **1996**, *29*, 761–782.
8. Preyer, S. Endoscopic ear surgery—A complement to microscopic ear surgery. *HNO* **2017**, *65*, 29–34. [CrossRef]
9. Tarabichi, M. Endoscopic transcanal middle ear surgery. *Indian J. Otolaryngol. Head Neck Surg.* **2010**, *62*, 6–24. [CrossRef]
10. Parelkar, K.; Thorawade, V.; Marfatia, H.; Shere, D. Endoscopic cartilage tympanoplasty: Full thickness and partial thickness tragal graft. *Braz. J. Otorhinolaryngol.* **2019**. [CrossRef]
11. Anschuetz, L.; Bonali, M.; Guarino, P.; Fabbri, F.B.; Alicandri-Ciufelli, M.; Villari, D.; Caversaccio, M.; Presutti, L. Management of Bleeding in Exclusive Endoscopic Ear Surgery: Pilot Clinical Experience. *Otolaryngol. Head Neck Surg.* **2017**, *157*, 700–706. [CrossRef]
12. Das, A.; Mitra, S.; Ghosh, D.; Kumar, S.; Sengupta, A. Does tranexamic acid improve intra-operative visualisation in endoscopic ear surgery? A double-blind, randomised, controlled trial. *J. Laryngol. Otol.* **2019**. [CrossRef] [PubMed]
13. Swarup, A.; le Nobel, G.J.; Andrysek, J.; James, A.L. The Current Limitations and Future Direction of Instrument Design for Totally Endoscopic Ear Surgery: A Needs Analysis Survey. *Otol. Neurotol.* **2018**, *39*, 778–784. [CrossRef]
14. Schuknecht, H.F.; Gulya, A.J. *Anatomy of the Temporal Bone with Surgical Implications*; Lea & Febiger: Philadelphia, PA, USA, 1986.
15. Badr-El-Dine, M.; James, A.L.; Panetti, G.; Marchioni, D.; Presutti, L.; Nogueira, J.F. Instrumentation and technologies in endoscopic ear surgery. *Otolaryngol. Clin. N. Am.* **2013**, *46*, 211–225. [CrossRef]
16. Kozin, E.D.; Kiringoda, R.; Lee, D.J. Incorporating Endoscopic Ear Surgery into Your Clinical Practice. *Otolaryngol. Clin. N. Am.* **2016**, *49*, 1237–1251. [CrossRef] [PubMed]
17. Mijovic, T.; Lea, J. Training and education in endoscopic ear surgery. *Curr. Otorhinolaryngol. Rep.* **2015**, *3*, 193–199. [CrossRef]
18. Alicandri-Ciufelli, M.; Molinari, G.; Beckmann, S.; Caversaccio, M.; Presutti, L.; Anschuetz, L. Epinephrine Use in Endoscopic Ear Surgery: Quantitative Safety Assessment. *ORL* **2019**. [CrossRef]
19. Zhang, Y.; Bai, Y.; Chen, M.; Zhou, Y.; Yu, X.; Zhou, H.; Chen, G. The safety and efficiency of intravenous administration of tranexamic acid in coronary artery bypass grafting (CABG): A meta-analysis of 28 randomized controlled trials. *BMC Anesthesiol.* **2019**, *19*, 104. [CrossRef] [PubMed]
20. Cao, G.; Chen, G.; Huang, Q.; Huang, Z.; Alexander, P.G.; Lin, H.; Xu, H.; Zhou, Z.; Pei, F. The efficacy and safety of tranexamic acid for reducing blood loss following simultaneous bilateral total knee arthroplasty: A multicenter retrospective study. *BMC Musculoskelet. Disord.* **2019**, *20*, 325. [CrossRef]
21. Ping, W.D.; Zhao, Q.M.; Sun, H.F.; Lu, H.S.; Li, F. Role of tranexamic acid in nasal surgery: A systemic review and meta-analysis of randomized control trial. *Medicine* **2019**, *98*, e15202. [CrossRef]
22. Gilroy, A.M.; MacPherson, B.R.; Ross, L.M.; Broman, J.; Josephson, A. *Atlas of Anatomy*; Thieme Stuttgart: Stuttgart, Germany, 2008.

23. Tomazic, P.V.; Hammer, G.P.; Gerstenberger, C.; Koele, W.; Stammberger, H. Heat development at nasal endoscopes' tips: Danger of tissue damage? A laboratory study. *Laryngoscope* **2012**, *122*, 1670–1673. [CrossRef]
24. Nelson, J.J.; Goyal, P. Temperature variations of nasal endoscopes. *Laryngoscope* **2011**, *121*, 273–278. [CrossRef] [PubMed]
25. MacKeith, S.A.; Frampton, S.; Pothier, D.D. Thermal properties of operative endoscopes used in otorhinolaryngology. *J. Laryngol. Otol.* **2008**, *122*, 711–714. [CrossRef] [PubMed]
26. Kozin, E.D.; Lehmann, A.; Carter, M.; Hight, E.; Cohen, M.; Nakajima, H.H.; Lee, D.J. Thermal effects of endoscopy in a human temporal bone model: Implications for endoscopic ear surgery. *Laryngoscope* **2014**, *124*, E332–E339. [CrossRef] [PubMed]
27. Aksoy, F.; Dogan, R.; Ozturan, O.; Eren, S.B.; Veyseller, B.; Gedik, O. Thermal effects of cold light sources used in otologic surgery. *Eur. Arch. Otorhinolaryngol.* **2015**, *272*, 2679–2687. [CrossRef]
28. Bottrill, I.; Perrault, D.F., Jr.; Poe, D. In vitro and in vivo determination of the thermal effect of middle ear endoscopy. *Laryngoscope* **1996**, *106*, 213–216. [CrossRef]
29. Craig, J.; Goyal, P. Insulating and cooling effects of nasal endoscope sheaths and irrigation. *Int. Forum Allergy Rhinol.* **2014**, *4*, 759–762. [CrossRef]

© 2020 by the authors. Licensee MDPI, Basel, Switzerland. This article is an open access article distributed under the terms and conditions of the Creative Commons Attribution (CC BY) license (http://creativecommons.org/licenses/by/4.0/).

Article
Surface Analysis of 3D (SLM) Co–Cr–W Dental Metallic Materials

Elena-Raluca Baciu [1], Ramona Cimpoeșu [2], Anca Vițalariu [1,*], Constantin Baciu [2], Nicanor Cimpoeșu [2,*], Alina Sodor [1], Georgeta Zegan [1] and Alice Murariu [1]

1. Faculty of Dental Medicine, "Grigore T. Popa" University of Medicine and Pharmacy, 700115 Iasi, Romania; elena.baciu@umfiasi.ro (E.-R.B.); alinasodor@yahoo.com (A.S.); georgeta.zegan@umfiasi.ro (G.Z.); alice.murariu@umfiasi.ro (A.M.)
2. Faculty of Materials Science and Engineering, "Gh. Asachi" Technical University, 700259 Iasi, Romania; ramona.cimpoesu@tuiasi.ro (R.C.); constantin.baciu@tuiasi.ro (C.B.)
* Correspondence: anca.vitalariu@umfiasi.ro (A.V.); nicanor.cimpoesu@tuiasi.ro (N.C.)

Featured Application: Specific prototyping of dental metal crowns and medical implants.

Abstract: The surface condition of the materials involved in dentistry is significant for the subsequent operations that are applied in oral cavity. Samples of Co–Cr–W alloy, obtained through selective laser melting (SLM) 3D printing, with different surface states were analyzed. Surface analysis after the 3D printing process and sandblasting was realized from microstructural, chemical composition, profilometry, droplet adhesion, scratch test, and microhardness perspectives. The results presented a hardening process and a roughness modification following the sandblasting procedure, a better adhesion of the liquid droplets, the appearance of micro-cracks during the scratch test, and the oxidation of the sample after the 3D printing process and surface processing.

Keywords: 3D printing; Co–Cr–W; dental materials; SEM; EDS; scratch test; droplet adhesion; profilometry; microhardness

1. Introduction

Characterized by a remarkable biocompatibility, cobalt-based alloys are widely used in the manufacturing of implants for the hip and knee, elements for immobilizing bone fractures (screws, plates, nails, etc.), and heart valves, and they have numerous applications in dentistry. These alloys can be: Binary (Co–Cr), ternary (Co–Cr–Mo and Co–Cr–Ni) or complex, with a content of approximately 65% Co, 30% Cr, and other elements (W, Ni, Ti, Nb, Si, etc.) in proportions of a maximum of 5%. These alloys belong to the category of stellites (lat. Stella = star), because they are characterized by a metallic luster persistent over time. They were originally produced by Deloro Stellite Company (Kokomo, IN, USA), based on the 1881 invention of Elwood Haynes [1–3]. Metal-based biomaterials are extremely important for improving and increasing the impact of implantable devices. Their applications mainly include load-bearing implants (e.g., metallic joints, such as hip), elements for stabilization, clips, wires, needles, plates, and screws, and a special category of dental implants and metal crowns proposed for the rehabilitation of deteriorated bone structure. Metal elements with applications in the field of medicine can be used as replacement elements for deteriorated tissues and in the recovery of soft tissues such blood vessels. The most commonly used metallic biomaterials are based on stainless steels, Co–Cr alloys, and Ti-based alloys. A new class of metal-based materials are biodegradable alloys, such as Mg-, Fe-, and Zn-based alloys. This special class can largely benefit from 3D printing technologies. Other examples of the alloys used in medical field are based on Ta, Nb, and noble materials such as Au- and Ag-based alloys with high percentages of Pt and Au. These materials are typically applied for the realization of dental prostheses such as crowns, dentures, inlays, and bridges [4,5].

Additive manufacturing (AM) was initially performed and developed in the 1980s as a new technique to improve the classical technologies for obtaining materials. In the case of metallic materials, the main property of these techniques (stereo-lithography, fused deposition model, selective electron beam melting, selective laser sintering, selective laser melting, and ink-jet printing) is based on the realization of one layer (different thicknesses) on top of another using a powder. The layers are connected through melting or fusing [6,7]. Based on the definition given by ASTM, the process of AM represents obtaining, layer by layer, metallic elements from a 3D design project as a bottom–top technique of making an element in a different way from the usual approach of melting [8–10]. From the different methods of obtaining dental materials based on Co–Cr–W (the tungsten element in Co–Cr alloy ensures increased tensile strength, increased hardness of alloys, and better resistance to corrosion and electro-corrosion [11–13]), additive manufacturing using selective laser melting equipment presents very good mechanical behavior [14]. The 3D printing of medical elements can address many of the specific difficulties in obtaining materials with special characteristics in the medical field, such as special structures of different thicknesses and, most importantly, 3D structures. Their surfaces after this CAD design process, at present, still require advanced surface processing, solutions for which are wanted at the industrial (mechanical or chemical) and laboratory (e.g., laser modifications of the surface) levels.

In the field of dentistry, the condition of the surface of metallic elements is very important for the activity undertaken in the oral environment in terms of the interaction between the metallic contact surface and the environment of existing biological solutions. The loss of dental material is higher in the case of higher surface exposure to a liquid. Among other reasons, the condition of the final surface of metallic materials for dentistry applications is important. The final goal of processing is to obtain a product whose execution quality is characterized by the dimensional accuracy of geometric shapes and the level of the surface profile.

This article analyzed the results obtained from the investigation of the surface condition of a metallic material (Co–Cr–W) obtained using 3D additive manufacturing technology (selective laser melting (SLM)). We propose different methods for the surface modification of dental materials that will improve the materials' behavior in an oral environment. After the CAD design of the final element, solidification was achieved by superimposing some metal powders in layers of 25 and 50 μm, respectively. The condition of the surface after its processing by two mechanical methods (sandblasting) was analyzed by scanning electron microscopy (SEM) in order to characterize the 2D and 3D profiles of the sample. Energy dispersive spectroscopy (EDS) chemical analysis was then conducted in order to characterize the chemical influence of the working processes on the surface. Additionally, profilometry was applied to characterize the roughness profile, while droplet adhesion tests were used in order to observe the hydrophobic character of the surface after the surface processing, and scratch and microhardness tests were carried out for surface hardening characterization.

2. Materials and Methods

2.1. Materials and Techniques

Complex dental problems (determined through computed tomography CT) need new solutions that involve custom items, created via the 3D printing process, in order to obtain specific elements (identified through 3D scanners) (Figure 1). The additive manufacturing technique (i.e., SLM) was used to obtain different samples from Co–Cr–W powders. High-purity metal powders of cobalt, chromium, and tungsten were used for the production through the SLM of the samples. The powder was provided by S&S Scheftner Germany and the 3D printing process was realized using SLM50 equipment (Realizer GmbH, Borchen/Paderborn, Germany) of the Romanian brand S.C. Lamas Microtech S.R.L. Bucharest.

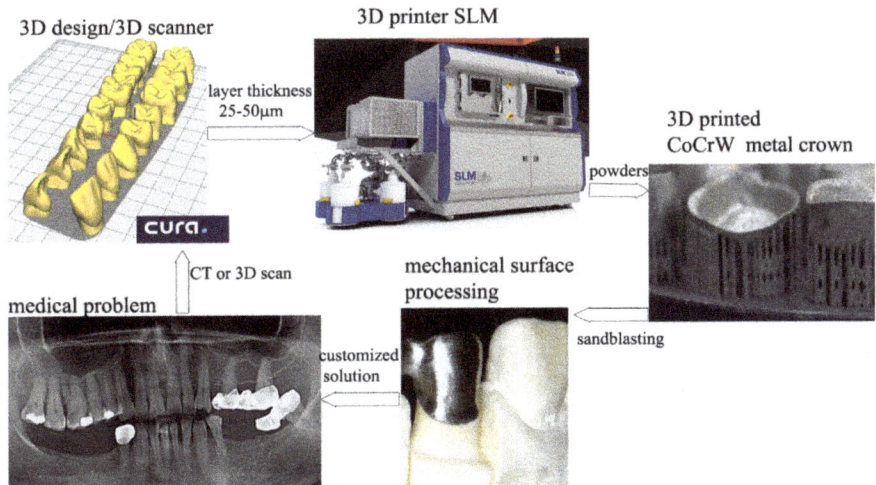

Figure 1. New solution for complex custom cases for dental configurations, a schematic set-up—with SLM.

The samples were made with different dimensions directly via the 3D printing process (for wettability determinations: 20 × 20 × 2 mm; scratch and profilometry: 30 × 10 × 2 mm; SEM and EDS were carried out before the main tests on the same samples).

Powders were added to the system of the 3D printing equipment, and the thickness of the layer was maintained at a constant value: t_{layer} = 25 and 50 µm. We obtained six experimental specimens by the additive manufacturing of Co–Cr–W alloy using a laser power of 70 W, a scanning rate of 1000 mm/s, and an exposure time 20 µs. The 3D printed samples had the follow chemical composition (average values from five different areas, wt%): Co: 59.38 (st. dev.: ±1.5), Cr: 25.99 (st. dev.: ±1.1), W: 9.54 (st. dev.: ±0.8), Mo: 3.48 (st. dev.: ±0.5), Si: max. 1.16 (st. dev.: ±0.1), and other elements: 0.43, determined through EDS analysis.

The samples were subject to mechanical preparation of the surfaces using two different sandblasting media. We noted the samples in the article, connected to the solidified layer during the 3D additive manufacturing process as: 25 µm—0 (non-sandblasted initial state), 25 µm—1 (sandblasted with Al_2O_3 particles), 25 µm—2 (successively sandblasted with Al_2O_3 and SiO_2 particles), which were similar for the 50 µm melted layer as well.

The surface modification of the 3D printed samples was realized by mechanical sandblasting with two different working particles. The sandblasting process was performed with specialized RenfertBasic Eco equipment (Renfert, Hilzingen, Germany), equipped with ceramic nozzles to determine the dimensions of the pulverized material. The alumina particles (Al_2O_3) had an F100 granulation (dimensions of 100–150 (±5) µm) and a working pressure of p = 3.5–4 bar. The silica particles had diameters between 77–110 (±5) µm and a working pressure of p = 2.5 bar. After they were polished, each sample was ultrasonically cleaned for 30 min in both distilled water and ethanol [15,16].

2.2. Wettability (Contact Angle) Determinations

The contact angles (θ_c) of electrolyte solution droplets (Hank electrolyte solution, CliniSciences, Nanterre, France) were determined at room temperature (20–25 °C) with drop shape analysis equipment (Kruss, Hamburg, Germany). The protractor was equipped with a KYOWA measuring system and a FAMAS interface analysis system. The camera located above the stage monitored the measuring point, and the moving camera measured the contact angle. The specially designed capillary (with an inner diameter of 5 µm) released a small drop in free fall. Five droplets of electrolyte solution were placed on the experimental metallic surfaces, and the resulting contact angles (θ_c in °) represent the averages of 10 measurements.

2.3. Microstructural and Chemical Surface Analysis

The surface of the samples was analyzed using a scanning electron microscope (SEM VegaTescan LMH II with an SE detector at 10 mm WD, 30 kV gun power supply, and VegaTC software for 2D and 3D images). The analysis was realized after the cleaning of the sample surface with N_2 flux. The chemical composition determinations were made using a Bruker EDS detector (5–6 kcps input signal, 15.5 mm working distance, and Esprit 2.2 software) on five different areas. Because the samples made from 25 and 50 μm layers presented similar results regarding the oxidation or contamination conditions, we have presented the average value of the elements percentages from both samples.

2.4. Profilometry, Scratch Test, and Microhardness

The profilometry was realized using Taylor Hobson FORM TALYSURF I50 equipment (tip: standard conical diamond; analyzed length: 30 mm; Talymap 3D Analysis software). We realized five determinations on each sample and have presented the average values.

Scratching resistance tests were realized using a CETR UMT-2 Tribometer. For the micro-scratch experiment, lamellas of 30 mm in length and 10 mm in width were used. A tip diamond indenter of the Rockwell type with a 120° opening angle at the edge was applied on the metallic 3D printed samples. The standard dimension values of Rockwell's indenter parameters are: Radius of 200 ± 10 μm; angle of 120° ± 0.35°; deviation from profile of ±2 μm. The software used for the micro-scratch test was the CETR-UMT Test Viewer.

The experiment was based on applying a variable force of 1–20 N over a distance of 35 mm with a 1 mm/s rate on a metallic surface. A normal force (F_z), a lateral force (F_x), and acoustic emission (AE) data were recorded with respect to the test time and the scratch distance during the test. The apparent coefficient of friction (COF) was calculated for each sample. These values are critical for accurately determining the sub-surface and incipient failure of the metallic materials.

A HVT-1000 micro Vickers hardness tester was used to analyze the microhardness of the samples (five determinations were made with a force of 300 gf (2.94 N), a test time of 15 s, and a distance between indentations of approximately 2.5 mm). The equipment had an JVC TK-C92 1EC optical camera and an eyepiece for fine adjustment of 40×.

3. Results and Discussion

The state of a metal surface is represented by a multitude of parameters: Roughness, defects induced by technological processing, microstructure, the size of the hardened surface layer (affected by internal stresses), physical–mechanical and chemical properties, etc. Surface quality has an important influence on the operating behavior and reliability of metal parts, regardless of their field of use.

3.1. Microstructural Surface State

It is known that the initial powder particles are mostly spherical, and the powder surface is clean and smooth [15]. Herein, the partial molten particles (Figure 2a,d) had more irregular shapes and their sizes were clearly bigger than those of the initial particles. After the evaluation of 100 particles prior to the SLM process, we obtained an average radius size of 12 μm, a minimum size of 6.1 μm, and a maximum size of 20.5, with a standard deviation of 2.6 μm. The circumference of the particles presented an average value of 75 μm (st. dev. 1.6) and a mean area of 468 μm² (st. dev.: 205). In the case of the semi-molten particles, after SLM, a union phenomenon was exhibited. The average size of the radius was 13.87 μm, with a minimum size of 8.85 μm, a maximum size of 38.63 μm, and a standard deviation of 0.86 μm (for the 25 μm—1 sample); meanwhile, the average size of the radius was 20.83 μm, with a minimum size of 8.41 μm, a maximum size of 89.31 μm, and a standard deviation of 15.27 μm (for the 50 μm—1 sample).

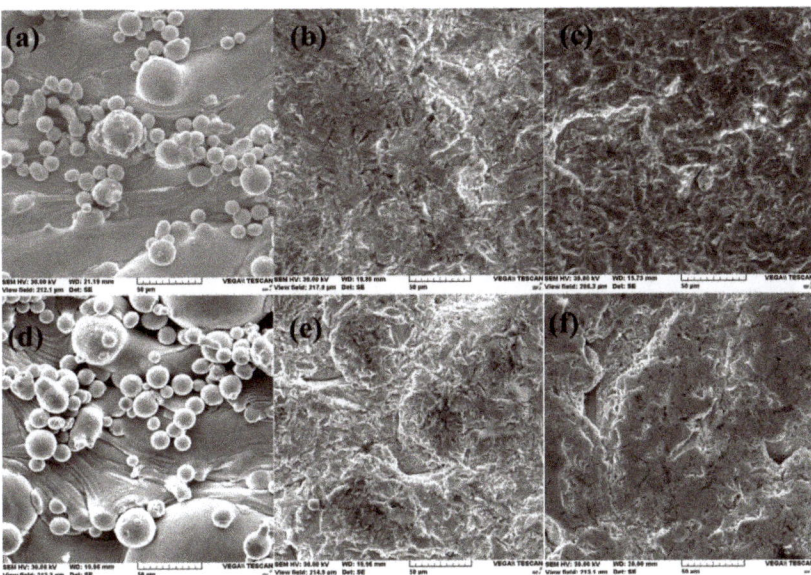

Figure 2. Scanning electron microscopy (SEM) (2D) images of the surface of the 3D printed samples at 1000×: (**a**) 25 μm—0, (**b**) 25 μm—1, (**c**) 25 μm—2, (**d**) 50 μm—0, (**e**) 50 μm—1, and (**f**) 50 μm—2.

The big dimensions and irregularities of the semi-melted particles in most of the printing strategies negatively impacted the powder melting and layer-by-layer coating [17].

This aspect, i.e., the surface of the material after 3D printing, is important for the selective laser melting process because this technique requires the melting of a powder layer (of a specific thickness) and recoating of this layer with a new one, up to hundreds of layers. The surface parameters of the melted powders must be strictly controlled due to their influence on the particle flux, which affects the laser consolidation [18]. For the semi-melted particles from the surface, a reduction in the number of attached parts of the particles can be observed in the main particles (Figure 2a,d).

After the sandblasting process, all of the particles remaining on the surface after SLM were removed (Figure 2b,c,e,f). The surfaces presented different roughnesses and aspects after the sandblasting process. The 3D images (Figure 3) confirm the results observed from the 2D images and present the aspects of the unified powders after SLM and the new roughnesses after the sandblasting process.

The structural and dimensional analyses show that the semi-melted particles remaining on the surface after the selective melting fabrication process are quite different from the initial powders in terms of size and appearance, and likely metallurgical characteristics as well, taking in consideration the high rate of cooling of the melted powder layers; however, no microstructural analyses were made in this article to confirm this.

3.2. Chemical Composition Determination

The chemical composition of the powders was determined through EDS: Co: 59; Cr: 25; W: 9.5; Mo: 3.50; Si: max. 1; other elements: max. 1.5 (wt%). Except for the presence of oxygen and carbon on the surface, no other elements were identified on the experimental materials after the SLM process (Table 1). The experimental results were obtained as averages of the values determined on five areas of 25 μm—0 (1, 2) and five areas of 50 μm—0 (1, 2). A greater oxidized surface was determined on the sample sandblasted only with alumina, while a smaller percentage of O was found on the surface of the sample processed with alumina and silica powders [19]. The growth and quality of a chromium oxide layer are extremely important for the high resistance to corrosion characteristics of Co–Cr-based alloys. The corrosion of Cr-based materials is based on the growth of a compact Cr oxide film (Cr_2O_3) or chromium-rich oxides, which act as a jam between the environment (oxidizing atmosphere) and the alloy. Frequently, the value of the corrosion resistance of these alloys depends on the properties of the Cr oxide formed [20].

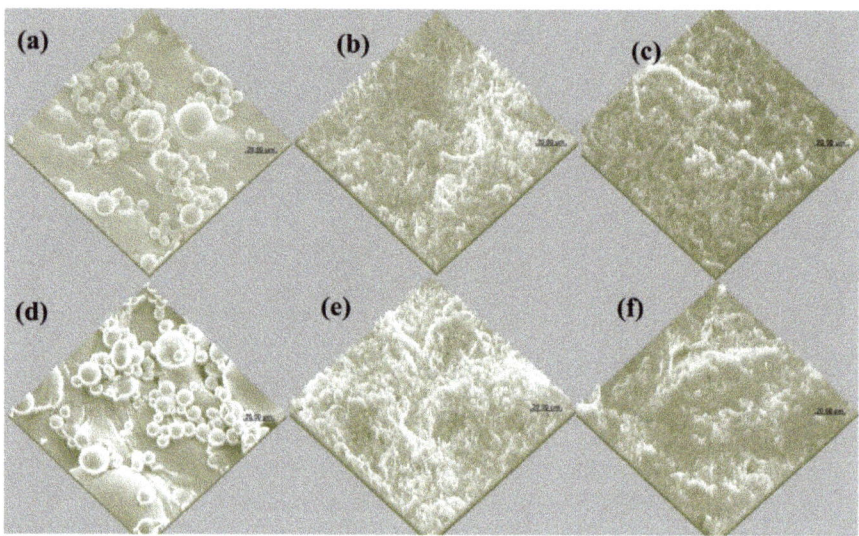

Figure 3. SEM (3D) images of the surface of the 3D printed samples at 1000×: (**a**) 25 μm—0, (**b**) 25 μm—1, (**c**) 25 μm—2, (**d**) 50 μm—0, (**e**) 50 μm—1, and (**f**) 50 μm—2.

Table 1. Chemical composition results of the experimental surface (average of 10 determinations).

Chemical Elements	Co		Cr		W		Mo		O		C		Al		Si	
	wt%	at%	wt%	at%	wt%	at%	wt%	at%	wt%	at%	wt%	at%	wt%	at%	wt%	at%
25 (50) μm—0	51.7	38.87	23.11	19.7	8.23	1.98	4.09	1.89	10.8	29.91	1.08	5.07	-	-	1.0	2.58
25 (50) μm—1	49.91	34.95	21.59	17.14	7.81	1.75	4.08	1.76	12.2	31.48	2.32	7.38	1.11	1.63	0.96	2.52
25 (50) μm—2	46.62	35.1	23.08	19.69	7.31	1.77	4.3	1.99	9.68	26.85	-	-	6.02	9.9	2.98	4.71
EDS error	1.35		0.66		0.3		0.31		2.9		0.5		0.2		0.2	

Standard deviation (SD; after 20 replicates of the chemical composition determination on the same 1 mm^2 area): Co: ±0.3; Cr: ±0.24; W: ±0.15; Mo: ±0.1; O: ±0.2; C: ±0.12; Al: ±0.1; Si: ±0.1.

After sandblasting with alumina and alumina with silica, the metallic surface was contaminated with alumina and silica parts. On the surface worked with $Al_2O_3+SiO_2$, no more carbon was identified based on the removal of the carbon compounds left behind by the SLM process (Figure 4).

The elemental distributions (made with an EDS detector on 25 μm—0 (1, 2)) of the main elements (Figure 4) determined using the experimental surfaces were in accordance with the chemical composition determinations presented in Table 1. All of the surfaces were covered with oxides and a reduced number of carbon-based compounds.

As a general observation in the literature, 3D printed materials typically present a certain oxidation state after the SLM process, and the surface is generally contaminated with particles of alumina and silica during the sandblasting process. A final cleaning stage using alcohol is recommended after the sandblasting operation.

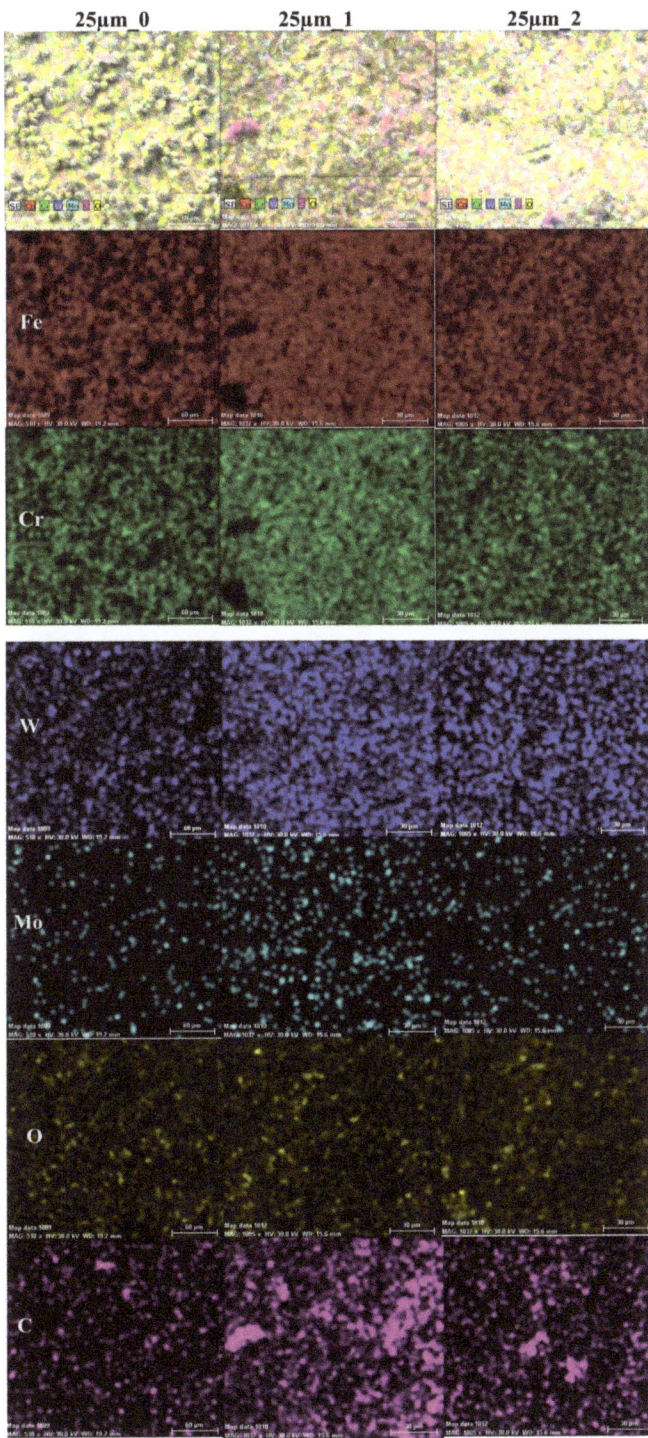

Figure 4. Elemental distributions (using an EDS detector) of the main elements determined on the experimental surfaces: All elements, Co, Cr, W, Mo, O, and C.

3.3. Profilometry

The state of a surface, for two materials placed in contact, is critical for the friction between them based on the specific comportment at the micro-scale of each metallic or non-metallic material. The surface profile is also important for the further deposition of a new material on top of the substrate.

The samples presented different profiles of the surface based on the processing mechanism applied (Figure 5). All of the samples exhibited a similar surface, with heights between 0.01 and 20 µm for most of them, and only a few peaks between 20 and 70 µm. The average roughness (Ra center line average (CLA)) was between 4.20 (sample 25 µm—2) and 8.77 µm (sample 50 µm—0), which influenced the friction coefficient, microhardness, and the wear resistance of the materials [21]. The average amplitude in the height direction (Rq) was also big for the 50 µm—0 (11.08 µm) sample and smallest for the 25 µm—2 (5.48 µm) sample, which had the smoothest surface (Table 2).

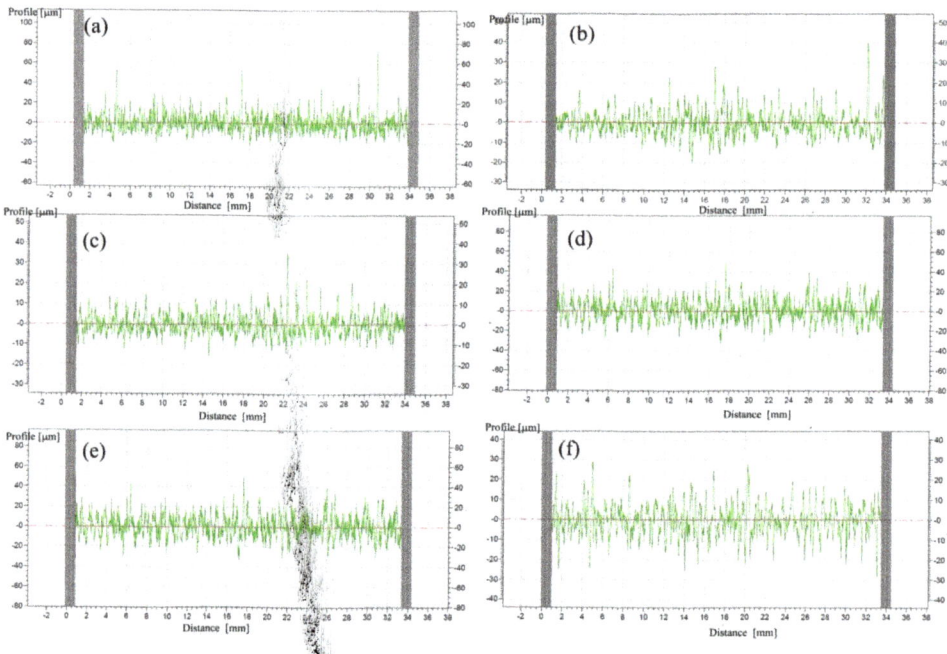

Figure 5. Profilometry of the surfaces of the samples: (**a**) 25 µm—0, (**b**) 25 µm—1, (**c**) 25 µm—2, (**d**) 50 µm—0, (**e**) 50 µm—1, and (**f**) 50 µm—2.

Table 2. The main profilometry parameters of the experimental elements.

Sample/Parameters	R_a µm	R_q µm	R_{sk}	R_{ku}	R_p µm	R_v µm	R_T µm
25 µm—0	7.16	9.73	1.57	9.06	36.65	18.30	94.53
25 µm—1	4.73	6.39	1.11	7.09	18.21	13.12	60.16
25 µm—2	4.20	5.48	0.90	5.69	16.62	12.35	51.64
50 µm—0	8.77	11.08	0.28	3.27	32.10	25.59	81.45
50 µm—1	8.00	10.05	0.07	3.05	23.61	24.54	68.82
50 µm—2	6.11	7.96	0.19	3.82	19.82	21.38	57.44

R_a, the arithmetic mean deviation of the profile or the average surface roughness; R_q, the mean square deviation of the profile; R_p, the maximum height, R_v, the maximum depth; R_T, the total height of the roughness profile (the difference between the height Zp of the highest peak and the depth of the deepest valley along the entire length of the assessment).

In order to evaluate a surface state in the height direction, two parameters were used: R_{sk} and R_{ku} (skewness and kurtosis). The value of the R_{sk} parameter indicates how a solid material is arranged as a profile of the surface and, in this case, all of the samples presented a positive skewness parameter, that is, the mean of the solid material was mostly below the mean line. The profiles characterized by high peaks, or at which the depression was gentle, had positive skewness parameter values.

Another characteristic of a surface is the kurtosis parameter (R_{ku}), which depends on the form of the profile. All of the surfaces of the samples had an R_{ku} parameter bigger than 3 (except the 50 µm—1 sample, which was R_{ku} = 3.05, meaning that the profile had relatively few high peaks and depressions, such as a platykurtic profile) (Table 2). This is the case of profiles with many high peaks and deep valleys (leptokurtic profile type). The Rp, Rv, and RT parameters confirm that the samples sandblasted with Al_2O_3 and SiO_2 presented the smoothest surface, and the biggest profile differences were found for the 25 µm—0 sample, while the smallest was found for the 25 µm—2 sample (the finest surface).

It is known that surfaces with a high roughness (lower quality) have a low fatigue resistance, while surfaces with a low roughness (high quality) have better fatigue resistance, as well as high resistance to corrosion. This does not mean that the over-finishing of surfaces, at the end of which a very low roughness is obtained, will always be beneficial to the normal operation of the parts and will not only generate an unjustified increase in their cost price.

There are numerous studies that have shown the dependence between the technological stages of processing and the quality of the surface obtained [22,23]. The electromechanical characteristics of the equipment used, the type of cutting tools used, the values of the working parameters adopted, and the nature of the coolants used represent specific technological factors that will exert their influence on the final roughness of machined parts. It is important to note that correlations can be made between the properties of the surface layer and the behavior of metal parts under the imposed operating conditions.

3.4. Wettability

Wettability is important because it accompanies all of the processes that take place at the gas–liquid–solid interfaces for all types of elements used in the medical field. If a metallic material is combined with a liquid, there is a tendency to form a protective liquid boundary layer that physically adheres to the surface of the metal. The protection also refers to the decrease in the corrosion intensity in the area of the boundary layer, which leads to a longer operation and to the preservation of the working dimensions of the functional parameters for a longer time. As the surface roughness increases (in the case of samples without mechanical surface processing), the solution droplets come into contact only with the surface peaks (its roughness), thereby decreasing the solid–liquid contact surface [24]. This phenomenon leads to the formation of an additional liquid–environment interface to complete the total liquid–metal interface, i.e., a decrease in the contact surface and implicitly of the metal corrosion [25]. The following average values of the contact angle for the samples were obtained from the determinations: 25 µm—0: 105° (±0.75°), 25 µm—1: 94° (±0.90°), and 25 µm—2: 72.25° (±1.05°); 50 µm—0: 112.5° (±0.70°), 50 µm—1: 101° (±0.90°), and 50 µm—2: 63.5° (±1.1°).

The experimental results presented in the literature have shown that an increased surface roughness decreases the dispersion percentage of liquid droplets [26]. The contact diffusion between the surface roughness and the experimental liquid droplet is shown in Figure 6, including the contact of a droplet with the surface of 25 (50) µm—0, which is a highly rough plane, of 25 (50) µm—1, which is a slightly rough plane, and of 25 (50) µm—2, which is a smooth plane.

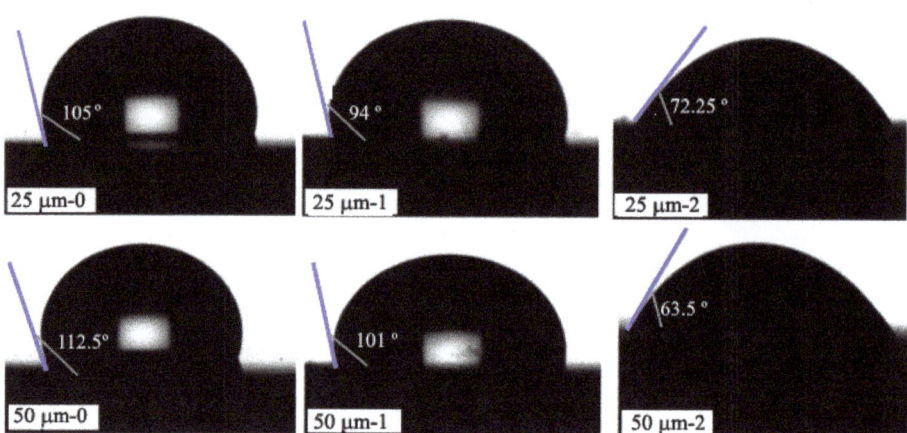

Figure 6. Wettability of the samples.

A smooth area will not realize any resistance toward a droplet, so it will quickly spread; in contrast, a rough area in contact with a droplet with the presence of air between the peaks will prevent dispersion. A higher roughness with more pits and valleys will store more air/voids and the layer of air will be stronger against liquid dispersion.

If the contact angle of the liquid with metal is high, the liquid will not form an adhesive film on the metal surface, no matter how fast the liquid moves (even at very low rates), so solid surfaces are exposed to both corrosion and erosion phenomena. The layer, depending on its thickness, protects the metal walls from the shock caused by solid particles that move with the biological liquid environment and also has a protective role for all types of dynamic aggression. Depending on the nature of the liquid with which it comes into contact (its chemical composition), corrosion phenomena may occur, but they will have a higher intensity than when combined with erosion phenomena. If the contact angle between the metal parts and the liquid (biological environment) is high, the liquid moistens the surface of the metal material and does not form a protective layer; meanwhile, if the contact angle is small, the liquid slides on the metal and forms a protective boundary layer [24].

For sanded surfaces with a low roughness (25 and 50 μm—2), the hydrophobic character of the surface at the electrolyte drops becomes hydrophilic (Figure 6), with the possibility of forming a protective boundary layer.

3.5. Hardness Analysis

The value of hardness is generally the value of the resistance of a material to the hard contact of another object. Hardness means the resistance of a material to cutting, friction, scratches, indentation, and permanent deformation [27,28]. This means that the surface of the material deforms permanently with a certain degree of impact.

The experimental setup in order to perform an increasing loading force for the scratch test was realized by a pin that was passed over the material surface with a linearly growing load up to the fracture point. On the surface, normal (F_z) and lateral (F_x) action were registered as functions of time or scratch distance (the tests were realized with 1 mm/s), pending the experiment. Along with these forces, AE data were registered and the coefficient of friction (COF) was calculated and plotted against the scratch length (mm). These parameters are important for the surface wear behavior and the initiation of material failure.

The F_x, COF, and AE were plotted against the scratch length (Figure 7) after performing the "in line" scratch test on the 3D printed samples.

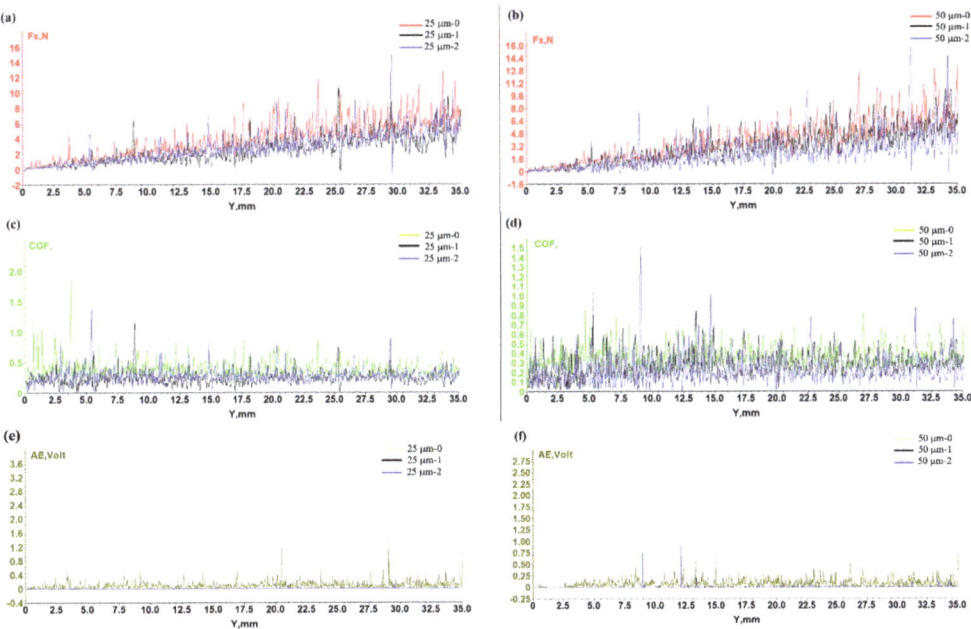

Figure 7. Scratch test results lateral force (F_x) (f) distance: (**a**) 25 μm and (**b**) 50 μm; coefficient of friction (COF) (f) distance: (**c**) 25 μm and (**d**) 50 μm; acoustic emission (AE) (f) distance: (**e**) 25 μm and (**f**) 50 μm.

The lateral force F_x applied to the samples presented a similar variation with distance behavior for all cases (Figure 7a,b), with slightly higher values for the 50 μm—0 (1, 2) samples based on their higher roughness. A few force peaks were observed for the 25 (50) μm—0 (1, 2) samples when harder surfaces were involved.

The apparent COF (can be extracted from: $F_x = COF\ F_z$) variations were similar for all of the samples, with a few peaks along the entire length of the scratch due to the particles being semi-melted and the presence of overlapped material Figure 7c,d. Slightly higher COF values for the initial samples were observed for both experimental sets. The determination of the COF and lateral force (F_x) indicated the presence of a critical load that was correlated with the AE signal. Here, the critical load represents the minimum load when the first crack occurs. In this case, there was no deposited layer to crack and the appearance of cracks was only locally determined for the initial samples by the deformation of semi-melted particles.

AE (occasionally named stress wave emission) presented variations over distance (Figure 7e,f) only for the initial state samples (25 (50) μm—0), where the samples had semi-melted powders (particles) and, during forced contact with the indentor, the variations in AE behavior represent the elastic waves generated during the release of energy internally stored in the material structure. When a scratch resistance limit of the surface through which it yields is reached, spontaneous explosions occur with known waves as a result of the formation and propagation of micro-cracks. Micro-fissures originate in surface notches when local stresses exceed the fracture stress of the material [29,30]. The heights of the AE evolution with distance represent micro-crack formation, and, on initial samples, most of them represent cracks and deformations of the semi-melted particles. For the sandblasted surface samples, only a few peaks of AE were observed in the case of the 50 μm—2 samples at 9 and 12.5 mm of the scratch length, and also at the end of the scratch (Figure 7f).

Based on the exceptional properties of the hardness and wear resistance, Co–Cr-based materials (ASTM F75, F90, F562, and F1537) are the proper choice for medical applications, where friction and sliding to other rough elements is necessary. Among biomedical materials, these ones present high hardness and very good wear resistance (300–400 HV) [27,31]. Nevertheless, the wear resistance characteristics of Co–Cr alloys strictly depend on their microstructure and the processes used to obtain/work them. For 3D printed samples, it is important that they behave as per their usual obtained alloy (made by casting) or even better in order to be used for similar applications.

The traces made on the surface of the material by scratch tests (Figure 8) present a deformation of the material under the F_z force with a flattening of semi-melted particles (Figure 8a,b) in the case of the initial surface. Moreover, micro-cracks were observed in the deformed powders on the scratch surface (Figure 8b), and the material was oriented in the scratch direction.

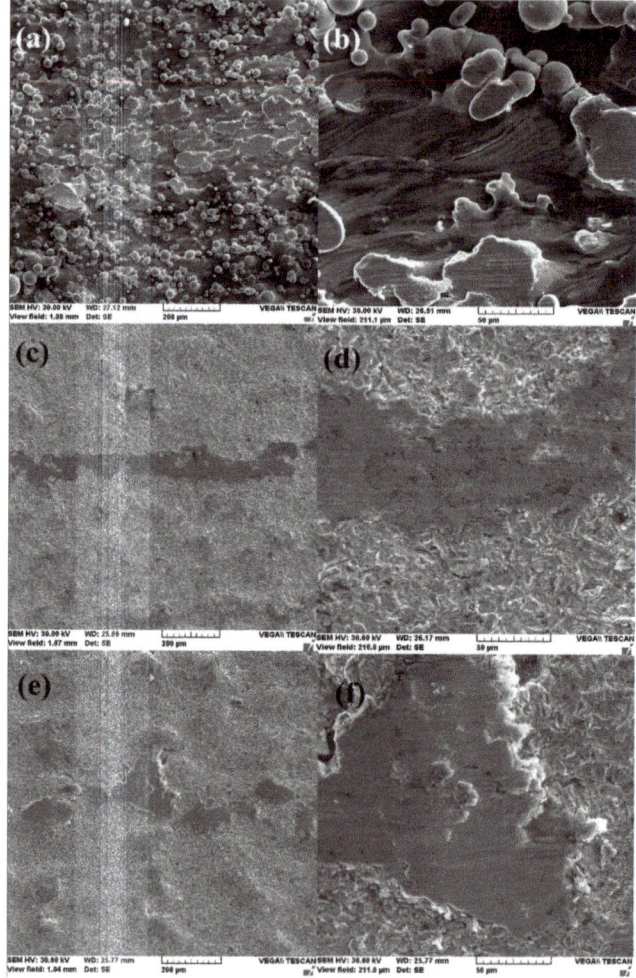

Figure 8. SEM images of the scratch surface of the (**a**) and (**b**) 50 µm—0, (**c**) and (**d**) 50 µm—1, and (**e**) and (**f**) 50 µm—2 for different amplification powers 500× and 1000×.

The scratch marks decreased as the width of the surface processing increased (Figure 8a,c,e) based on the material response because of the hardening phenomenon, which accompanied the sandblast process. No micro-cracks were observed for the smoother 25 (50) µm—1 and —2 samples (Figure 8d,f) except at the end of the scratch of the 50 µm—2 sample, which confirms the AE results from Figure 7f. For both sandblasted samples, the scratch marks presented interruptions, caused by the surface profilometry and hardness. Material overlap was observed on the surface of the 50 µm—2 samples at higher scratch forces (at the end of the test with a force of 20 N).

The Vickers microhardness results (HV) (Table 3) show a hardening of the surface with the sandblast process in both samples sets. Liu and others identified microhardness (2 N/15 s) of the Co–Cr–W after different heat treatments between 350 and 550 HV [32]. The hardness values obtained from five different locations highlight the differences between the initial surface and the sandblasted surfaces, especially in the case of the first set of samples (25 µm—0 (1, 2)). Different values of the micro Vickers hardness on the same surface were observed for all of the samples based on the presence of semi-melted initial particles and different hardened areas. The values obtained recommend these materials for dental applications as metal crowns [33].

Table 3. Micro Vickers hardness of the samples.

Sample	Test 1	Test 2	Test 3	Test 4	Test 5	Average HV Value
25 µm—0	404.3	406.1	437.5	378	411.7	407.52
25 µm—1	391.7	470.4	456.8	468	429	443.18
25 µm—2	479.7	499.4	528.4	514	475	499.3
50 µm—0	468	437.5	463.5	472.7	463.5	461.04
50 µm—1	459	472.7	484.5	470.4	472.7	471.86
50 µm—2	477.4	470.4	481.2	484.5	473.8	477.46

On the initial 25 µm—0 sample, we observed a value much lower than the other four (378 HV), which was probably taken from a semi-melted particle on the surface. In the case of the 25 µm samples, a high increase in the microhardness was observed with the application of sandblasting, especially for $Al_2O_3+SiO_2$ solicitation, which supplementarily hardens the surface. For samples made of thicker layers (50 µm), the surface hardening presented smaller differences to the first set (25 µm), which exhibited average values between the minimum and maximum values. The variation of the microhardness on the material surface confirmed the results obtained from scratch tests. Comparing the results of the hardness for the average values, an almost 23% increase was determined for the first set of samples and 3.5% for the second set.

4. Conclusions

Analyzing the results obtained from the surface investigation of some 3D printed metallic samples, we present the follow conclusions:

- After the 3D printing process, the dental material samples made through SLM presented many semi-melted particles that increased the contact surface of the elements. Using different blasting media, the 3D printed alloy surface could be cleaned and its roughness modified in order to fulfill the medical application requirements.
- The oxidation process took place on the surface of all of the samples, and unstable compounds based on sandblast materials were identified.
- The 50 µm—0 (1, 2) presented a higher roughness than the 25 µm—0 (1, 2) sample for the Ra, Rq, and RT parameters, regardless of the condition of the samples' surfaces.
- The droplet adhesion characteristics were modified with the sandblast operation.
- The sandblast operation considerably increased the samples' hardness, especially in the case of 25 µm—0 (1, 2).
- Even the material is consecrate in literature, and, in laboratory practice, further medical tests can be considered in order to confirm the implications of surface preparation, especially on 3D complex structures.

Author Contributions: Conceptualization, E.-R.B., C.B., and N.C.; methodology, R.C. and G.Z.; software, R.C. and A.S.; validation, A.M. and A.V.; investigation, N.C., C.B., and E.-R.B.; resources, R.C.; writing—original draft preparation, N.C.; writing—review and editing, R.C. and G.Z.; visualization, A.M.; supervision, N.C. All authors read and agreed to the published version of the manuscript.

Funding: This research was funded by "Grigore T. Popa" University of Medicine and Pharmacy Iași grant number 27500, "Influence of 3D printing/Selective Laser Melting processing parameters on surface quality and bonding level between metal/ceramic components in dental prostheses." And The APC was funded by "Grigore T. Popa" University of Medicine and Pharmacy Iași.

Institutional Review Board Statement: Not applicable.

Informed Consent Statement: Not applicable.

Acknowledgments: This work was supported by a young-researcher grant from "Grigore T. Popa" University of Medicine and Pharmacy Iași, project number 27500, "Influence of 3D printing/ Selective Laser Melting processing parameters on surface quality and bonding level between metal/ceramic components in dental prostheses."

Conflicts of Interest: The authors declare no conflict of interest.

References

1. Niinomi, M.; Narushima, T.; Nakai, M. Advances in Metallic Biomaterials, Tissues, Materials and Biological Reactions. In *Springer Series in Biomaterials Science and Engineering*; Springer: Berlin/Heidelberg, Germany, 2015; Volume 3, pp. 157–178.
2. Wulfman, C.; Koenig, V.; Mainjot, A.K. Wear measurement of dental tissues and materials in clinical studies: A systematic review. *Dent. Mater.* **2018**, *34*, 825–850. [CrossRef] [PubMed]
3. Eakle, S.W.; Kimberly, B. Dental Materials. In *Clinical Applications for Dental Assistants and Dental Hygienists, Saunders*, 4th ed.; Elsevier: Amsterdam, The Netherlands, 2020; pp. 220–236.
4. Zeng, L.; Xiang, N.; Wei, B. A comparison of corrosion resistance of cobalt-chromium-molybdenum metal ceramic alloy fabricated with selective laser melting and traditional processing. *J. Prosthet. Dent.* **2014**, *112*, 1217–1224. [CrossRef] [PubMed]
5. Takaichi, A.; Suyalatu, N.T.; Natsuka, J.; Naoyuky, N.; Yusuke, T.; Satoshi, M.; Doi, H.; Kurosu, S.; Chiba, A.; Wakabayashi, N.; et al. Microstructures and mechanical properties of Co-29Cr-6Mo alloy fabricated by selective laser melting for dental application. *J. Mech. Behav. Biomed. Mater.* **2013**, *21*, 67–76. [CrossRef] [PubMed]
6. Van Noort, R. The future of dental devices is digital. *Dent. Mater.* **2012**, *28*, 3–12. [CrossRef]
7. Dikova, T.; Dzhendov, D.; Simov, M.; Katreva-Bozukova, I.; Angelova, S.; Pavlova, D.; Abadzhiev, M.; Tonchev, T. Modern trends in the development of the technologies for production of dental constructions. *J. IMAB* **2015**, *21*, 974–981. [CrossRef]
8. Gu, D. *Laser Additive Manufacturing of High-Performance*; Springer-Verlag: Berlin/Heidelberg, Germany, 2015; pp. 15–71.
9. Bandyopadhyay, A.; Bose, S.; Das, S. 3D printing of biomaterials. *MRS Bulletin* **2015**, *40*, 108–115. [CrossRef]
10. Torabi, K.; Farjood, E.; Hamedani, S.H. Rapid prototyping technologies and their applications in prosthodontics, a review of literature. *J. Dent.* **2015**, *16*, 1–9.
11. Lu, Y.; Lin, W.; Xie, M.; Xu, W.; Liu, Y.; Lin, J.; Yu, C.; Tang, K.; Liu, W.; Yang, K.; et al. Examining Cu content contribution to changes in oxide layer formed on selective-laser-melted CoCrW alloys. *Appl. Surf. Sci.* **2019**, *464*, 262–272. [CrossRef]
12. Wang, D.; Ye, G.; Dou, W.; Zhang, M.; Yang, Y.; Mai, S.; Liu, Y. Influence of spatter particles contamination on densification behavior and tensile properties of CoCrW manufactured by selective laser melting. *Opt. Laser Technol.* **2020**, *121*, 105678. [CrossRef]
13. Ho, L.W.; Jung, K.-H.; Hwang, S.-K.; Kang, S.-H.; Kim, D.-K. Microstructure and mechanical anisotropy of CoCrW alloy processed by selective laser melting. *Mater. Sci. Eng. A* **2019**, *749*, 65–73.
14. Padrós, R.; Punset, M.; Molmeneu, M.; Velasco, A.B.; Herrero-Climent, M.; Rupérez, E.; Gil, F.J. Mechanical Properties of CoCr Dental-Prosthesis Restorations Made by Three Manufacturing Processes Influence of the Microstructure and Topography. *Metals* **2020**, *10*, 788.
15. Baciu, A.M.; Bejinariu, C.; Corăbieru, A.; Mihalache, E.; Lupu-Poliac, M.; Baciu, C.; Baciu, E.R. Influence of process parameters for Selective Laser Melting on the roughness of 3D printed surfaces in Co-Cr dental alloy powder. *IOP Conf. Ser. Mater. Sci. Eng.* **2019**, *572*, 012054. [CrossRef]
16. Bernevig-Sava, M.A.; Stamate, C.; Lohan, N.-M.; Baciu, A.M.; Postolache, I.; Baciu, C.; Baciu, E.-R. Considerations on the surface roughness of SLM processed metal parts and the effects of subsequent sandblasting. *IOP Conf. Ser. Mater. Sci. Eng.* **2019**, *572*, 012071. [CrossRef]
17. Spierings, A.B.; Herres, N.; Levy, G. Influence of the particle size distribution on surface quality and mechanical properties in AM steel parts. *Rapid Prototyp. J.* **2011**, *17*, 195–202. [CrossRef]
18. Tan, J.H.; Wong, W.L.E.; Dalgarno, K.W. An overview of powder granulometry on feedstock and part performance in the selective laser melting process. *Addit. Manuf.* **2017**, *18*, 228–255. [CrossRef]
19. Wang, D.; Wu, S.; Fu, F.; Mai, S.; Yang, Y.; Liu, Y.; Song, C. Mechanisms and characteristics of spatter generation in SLM processing and its effect on the properties. *Mater. Des.* **2017**, *117*, 121–130. [CrossRef]

20. Dorcheh, A.S.; Schütze, M.; Galetz, C.M. Factors affecting isothermal oxidation of pure chromium in air. *Corros. Sci.* **2018**, *130*, 261–269. [CrossRef]
21. Hong, M.-H.; Min, B.K.; Kwon, T.-Y. The Influence of Process Parameters on the Surface Roughness of a 3D-Printed Co–Cr Dental Alloy Produced via Selective Laser Melting. *Appl. Sci.* **2016**, *6*, 401. [CrossRef]
22. Wei, C.; Luo, L.; Wu, Z.; Zhang, J.; Su, S.; Zhan, Y. New Zr-25Ti-xMo alloys for dental implant application: Properties characterization and surface analysis. *J. Mech. Behav. Biomed.* **2020**, *111*, 104017.
23. Pasang, T.; Lees, S.; Takahashi, M.; Fujita, T.; Conor, P.; Tanaka, K.; Kamiya, O. Machining of dental Alloys: Evaluating the surface finish of laterally milled Co-Cr-Mo Alloy. *Procedia Manuf.* **2017**, *13*, 5–12. [CrossRef]
24. Qin, L.; Wu, H.; Guo, J.; Feng, X.; Dong, G.; Shao, J.; Zeng, Q.; Zhang, Y.; Qin, Y. Fabricating hierarchical micro and nano structures on implantable Co-Cr-Mo alloy for tissue engineering by one-step laser ablation. *Colloid Surf. B.* **2018**, *161*, 628–635. [CrossRef] [PubMed]
25. Zhang, Z.Y.; Gu, Q.M.; Jiang, W.; Zhu, H.; Xu, K.; Ren, Y.P.; Xu, C. Achieving of bionic super-hydrophobicity by electrodepositing nano-Ni-pyramids on the picosecond laser-ablated micro-Cu-cone surface. *Surf. Coat. Technol.* **2019**, *363*, 170–178. [CrossRef]
26. Rupp, F.; Gittens Rolando, A.; Lutz, S.; Marmur, A.; Boyan, B.D.; Schwartz, Z.; Geis-Gerstorfer, J. A review on the wettability of dental implant surfaces I: Theoretical and experimental aspects. *Acta Biomater.* **2014**, *10*, 2894–2906. [CrossRef] [PubMed]
27. Kuncická, L.; Kocich, R.; Lowe, T.C. Advances in metals and alloys for joint replacement. *Prog. Mater. Sci.* **2017**, *88*, 232–280. [CrossRef]
28. Hong, J.H.; Yeoh, F.Y. Mechanical properties and corrosion resistance of cobalt-chrome alloy fabricated using additive manufacturing. *Mater. Today Proc.* **2020**, *29*, 196–201. [CrossRef]
29. McGrory, B.J.; Ruterbories, J.M.; Pawar, V.D.; Thomas, R.K.; Salehi, A.B. Comparison of surface characteristics of retrieved cobalt-chromium femoral heads with and without ion implantation. *J. Arthroplast.* **2012**, *27*, 109–115. [CrossRef]
30. Ratner, B.D.; Hoffman, A.S.; Schoen, F.J.; Lemons, J.E. *Biomaterials Science: An Introduction to Materials in Medicine*; Academic Press: Cambridge, MA, USA, 2004.
31. Paleu, V.; Gurau, G.; Comaneci, R.I.; Sampath, V.; Gurau, C.; Bujoreanu, L.G. A new application of Fe-28Mn-6Si-5Cr (mass%) shape memory alloy, for self-adjustable axial preloading of ball bearings. *Smart Mater. Struct.* **2018**, *27*, 075026. [CrossRef]
32. Lu, Y.; Zhao, W.; Yang, C.; Liu, Y.; Xiang, H.; Yang, K. Improving mechanical properties of selective laser melted Co29Cr9W3Cu alloy by eliminating mesh-like random high-angle grain boundary. *Mater. Sci. Eng. A* **2020**, *793*, 139895. [CrossRef]
33. Cicciu, M.; Fiorillo, L.; D'Amico, C.; Gambino, D.; Amantia, E.M.; Laino, L.; Crimi, S.; Campagna, P.; Bianch, A.; Herford, A.S.; et al. 3D Digital Impression Systems Compared with Traditional Techniques in Dentistry: A Recent Data Systematic Review. *Materials* **2020**, *13*, 1982. [CrossRef]

Article

Finite Element Analysis of Mandibular Anterior Teeth with Healthy, but Reduced Periodontium

Ioana-Andreea Sioustis [1], Mihai Axinte [2,*], Marius Prelipceanu [3,*], Alexandra Martu [1,*], Diana-Cristala Kappenberg-Nitescu [1], Silvia Teslaru [1], Ionut Luchian [1,†], Sorina Mihaela Solomon [1,†], Nicanor Cimpoesu [2] and Silvia Martu [1]

1. Periodontology Department, Faculty of Dentistry, "Grigore T. Popa" University of Medicine and Pharmacy of Iasi, 700115 Iasi, Romania; ioana-andreea.sioustis@umfiasi.ro (I.-A.S.); diana-cristala.nitescu@umfiasi.ro (D.-C.K.-N.); silvia.teslaru@umfiasi.ro (S.T.); ionut.luchian@umfiasi.ro (I.L.); sorina.solomon@umfiasi.ro (S.M.S.); silvia.martu@umfiasi.ro (S.M.)
2. Materials Science Department, Faculty of Materials Science and Engineering, Technical University "Gheorghe Asachi" of Iasi, 700050 Iași, Romania; nicanor.cimpoesu@tuiasi.ro
3. Integrated Center for Research, Development and Innovation in Advanced Materials, Nanotechnologies, and Distributed Systems for Fabrication and Control, Department of Computers, Electronics and Automation, Ștefan cel Mare University of Suceava, 720229 Suceava, Romania
* Correspondence: mihai.axinte@academic.tuiasi.ro (M.A.); marius.prelipceanu@usm.ro (M.P.); maria-alexandra.martu@umfiasi.ro (A.M.)
† Authors with equal contribution as the first author.

Citation: Sioustis, I.-A.; Axinte, M.; Prelipceanu, M.; Martu, A.; Kappenberg-Nitescu, D.-C.; Teslaru, S.; Luchian, I.; Solomon, S.M.; Cimpoesu, N.; Martu, S. Finite Element Analysis of Mandibular Anterior Teeth with Healthy, but Reduced Periodontium. *Appl. Sci.* **2021**, *11*, 3824. https://doi.org/10.3390/app11093824

Academic Editor: Gianrico Spagnuolo

Received: 30 March 2021
Accepted: 20 April 2021
Published: 23 April 2021

Publisher's Note: MDPI stays neutral with regard to jurisdictional claims in published maps and institutional affiliations.

Copyright: © 2021 by the authors. Licensee MDPI, Basel, Switzerland. This article is an open access article distributed under the terms and conditions of the Creative Commons Attribution (CC BY) license (https://creativecommons.org/licenses/by/4.0/).

Abstract: Finite element analysis studies have been of interest in the field of orthodontics and this is due to the ability to study the stress in the bone, periodontal ligament (PDL), teeth and the displacement in the bone by using this method. Our study aimed to present a method that determines the effect of applying orthodontic forces in bodily direction on a healthy and reduced periodontium and to demonstrate the utility of finite element analysis. Using the cone-beam computed tomography (CBCT) of a patient with a healthy and reduced periodontium, we modeled the geometric construction of the contour of the elements necessary for the study. Afterwards, we applied a force of 1 N and a force of 0.8 N in order to achieve bodily movement and to analyze the stress in the bone, in the periodontal ligament and the absolute displacement. The analysis of the applied forces showed that a minimal ligament thickness is correlated with the highest value of the maximum stress in the PDL and a decreased displacement. This confirms the results obtained in previous clinical practice, confirming the validity of the simulation. During orthodontic tooth movement, the morphology of the teeth and of the periodontium should be taken into account. The effect of orthodontic forces on a particular anatomy could be studied using FEA, a method that provides real data. This is necessary for proper treatment planning and its particularization depends on the patient's particular situation.

Keywords: finite element analysis; orthodontics; healthy and reduced periodontium; bodily movement; mandibular anterior teeth

1. Introduction

Through the ability to investigate or simulate correctly and accurately a series of medical phenomena, modern, computerized technology has become a necessity, offering the advantage of efficiency in the medical act, in a minimally invasive manner for the patient. Used for a long time in biomechanics to study physical phenomena in two-dimensional and three-dimensional spaces, the finite element method has become a tool increasingly used in medicine. By simulating dental movements and various stresses that are exerted on the teeth or on the periodontium during orthodontic treatment, FEM offers the advantage of correct prediction and appropriate therapy [1].

Recent studies have signaled the opportunity to use this method of investigation both in orthodontics and in implantology or prosthetic rehabilitation [1,2]. This justifies the interest shown by specialists in the field of dentistry to achieve mathematical models capable of simulating the evolution of medical processes such as stress variation in the enamel and dentin in restorative dentistry [3], assessment of implant prosthetic rehabilitation in mandibular bone atrophy [4], prediction of the endodontic treatment outcome [5], finite element study during orthodontic treatment [6] and stress in the periodontal ligament as a result of orthodontic treatment in association with periodontal disease [7]. Finite element analysis is a useful method with effective results in determining the effects of orthodontic forces during orthodontic treatment. This method is used to determine the quantitative mechanical stress acting on a biological structure under the assigned material properties (FEA), whereas it is the only method that can provide stress distribution data. However, it can also provide sensitive data, such as stress distribution, data that are provided only through this method.

During orthodontic tooth movement, the applied force initiates the cell processes occurring in dental and periodontal tissues [8], thus resulting in a cellular response that makes the displacement of the tooth possible [9].

To the best of our knowledge, this is the first study in which the entire finite element modeling and construction of a healthy and reduced periodontium of anterior mandibular teeth was constructed using the protocol described below. The aim of the study was to analyze the effect of the application of orthodontic forces in bodily direction on a healthy and reduced periodontium and to demonstrate the utility of FE analysis. This is achieved through a correlation between in vivo situations and FEA results. Therefore, the FEA method and the object-oriented programming used in this study are very useful for establishing a personalized treatment plan with specific steps depending on the patient's particular periodontal and dental anatomy.

2. Materials and Methods

To perform this study, we used images from the cone-beam computed tomography (CBCT) of a patient with a healthy and reduced periodontium using a Planmeca ProMax® 3D Mid (Helsinki, Finland). The CBCT's settings were the following: mode (11.975 s, 674 mGy·cm^2, 85 kV, and 11 mA); field of view (FOV), 80 × 50 mm, the lower facial third, specifically the anterior mandibular area; and voxel size, 0.2 mm. After analyzing the sections of the mandible, we made the geometric construction of the contour of the elements necessary for the study. We saved 57 consecutive sections, at equal distance from the CBCT.

The stages of constructing the model for the bone structure were the following: first, 57 consecutive sections were saved (at equal distance from the CBCT); measurements of the horizontal distance between two landmarks in the image were performed using the Planmeca Romexis® software and CatiaV5 R20 program. The distance between two consecutive horizontal planes was given directly by the CBCT and resulted from the Planmeca Romexis® software. After performing these operations, 57 sectioning planes, meaning 56 intervals (d_t = total distance n_{interv} = number of intervals (spaces between two consecutive planes), interv d_t = 413, interv = d_t/n_{interv} = 413/59 = 7 mm) were obtained. Next, the distance between two CBCT sections was obtained. These data were used in Catia V5R20 to build the planes on which the images were later placed; in the foreground, a sketch with the size of the image's resolution (1920 × 1080) was made and the first image was introduced. In order to accomplish this, the *texture* option in the material parameterization window was utilized. In this way, 1 pixel was represented by 1 mm. Next, the position on the plane was established by approximation, on a large magnification scale, using marks from the image introduced on the first plane. The positioning was done in relation to the three teeth already modeled and to the images with the sections on the basis of which they were made. The measurements were performed using both the Planmeca Romexis® software and the CatiaV5 R20 program in order to accomplish the following: initially, to transfer the real data from the CBCT to the CatiaV5 R20 program, and afterwards, during the stages of modeling, to verify the reproduction's correctness and the most accurate replication of the data. Thus, the repeatability and precision of the described methodology were ensured.

A. The steps involved in producing the 57 plans with the CBCT scan images were as follows:
1. The construction of the CBCT image placement plans. In the Catia V5R20 program, 57 planes parallel to each other were built with the help of the *plane* tool, at a distance of 7 mm from each other. Then, on each plane, a frame with the size 1920 × 1080, equal to the resolution of the image saved in the Planmeca Romexis® Viewer program, was inserted (Figure 1).

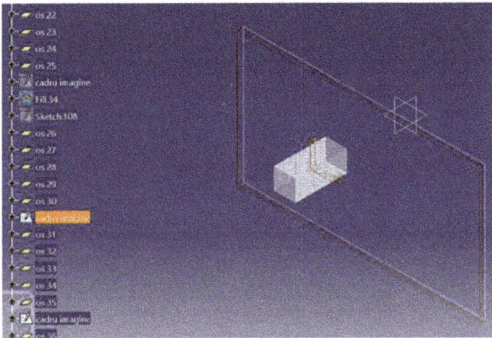

Figure 1. The insertion of a frame on each plane that was previously created.

2. The plans were constructed with CBCT images, in the GSD (Geometrical Shape Design) module of the CatiaV5 R 20 program. On each frame, introduced in the previous step, the corresponding image saved from the CBCT was introduced; after this step, the frame was transformed into a surface in the GSD module of the Catia V5R20 program. Subsequently, a material was assigned to each surface.
3. The saved image from the CBCT was applied and was set to the selected plane. This was made with the help of the "texture" section of the "properties" menu of the flat surface on which the CT image was applied.
4. The required outline was copied. At this stage, the outer contour of the mandibular bone structure was copied for each section. Copying was done manually using the *spline* function from the *Part design* module, from *Sketch tools*, in Catia V5R20. It is known that the result and quality of this step, the step of making the contour of the section through the bone, depends on the operator's experience.
5. The next step included fixing and "constraining" the obtained contour. This step is mandatory for each 2D entity in the sections obtained. The constraint aims to fix the points introduced to create curves. If this step is not performed, there is a high risk of error. This step was performed by using the *fixed together* tool from the *Constraints* menu. A dimensional constraint was also applied, with the aim of fixing the position of all points in relation to the origin of the work plane (Figure 2).
6. Rearranging the operations in the operations list of the Catia V5 R20 program. This step is necessary in order to easily establish a correlation between sections to the subsequent steps of obtaining the 3D model. It was done by moving the corresponding branches one by one, for each of the 57 sections.
7. Hiding the construction elements used in Catia V5 R20. This is necessary in order to make the contour for the stage of obtaining the 3D model. All the introduced elements (construction plans, frames and images from the sections) were hidden in order to be able to work continuously with the obtained sections. This was done for each plane separately.

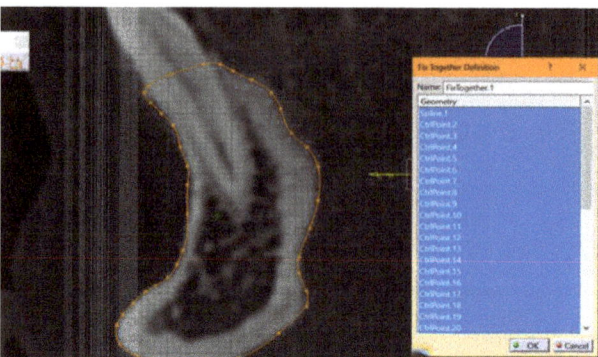

Figure 2. Fixing and "constraining" the position of contour points in sections.

8. Obtaining the 3D model of the mandible section. The construction of the mandible was performed based on the sections from Figure 3.

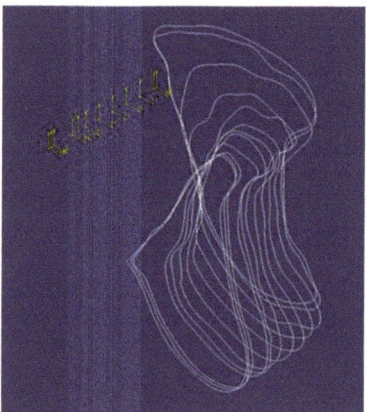

Figure 3. The 2D wireframe sections, necessary to make the 3D model of the mandible.

9. Converting 2D sections. The "*Multi-sections surface*" 3D construction tool was used to convert the previously built 2D sections, as seen in Figure 4. The orientation of the curves was taken into account so that they all had the same tangent direction at the starting point. These were the starting points for the construction of 2D sections, for each section. For this reason, the construction of these sections is particularly important. The order of selection was also taken into account, as shown in the window in Figure 4.

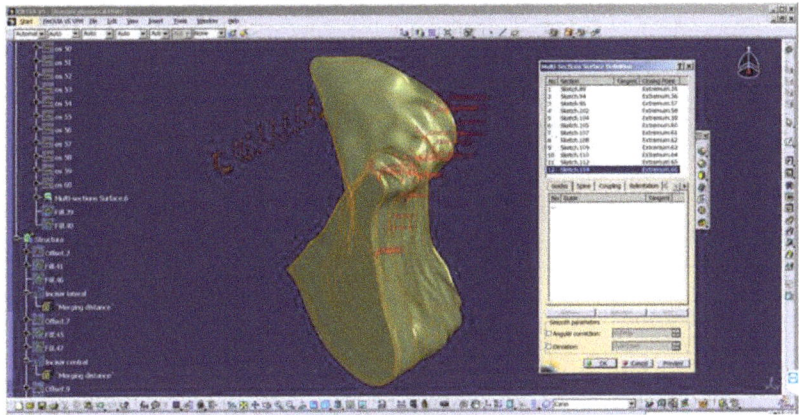

Figure 4. Converting 2D sections to 3D model.

10. The final structure of the mandible. After the construction of the 3D model of the mandibular section, the 2D structure of the sections was hidden.

B. Obtaining the 3D model of the teeth according to the CBCT scan model.

The same operations as for the bone structure were performed in order to obtain the 3D model of the teeth. Each tooth was modeled separately, following the same steps as above. Figure 5 shows the construction of the canine's geometry and Figure 6 shows the model obtained for three teeth (central incisor, lateral incisor and canine).

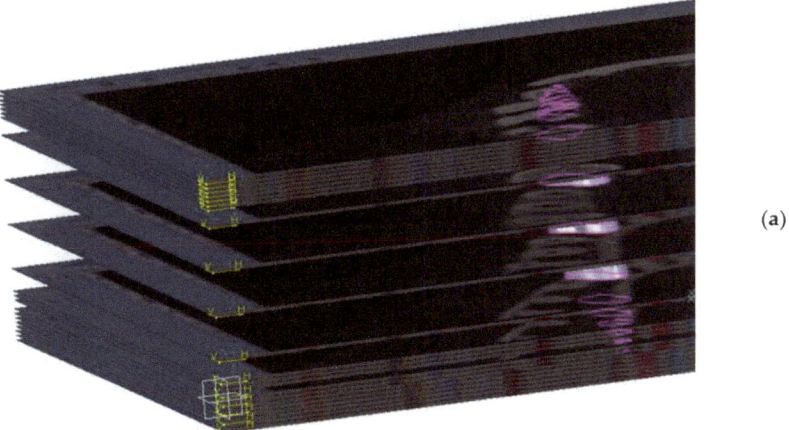

(a)

Figure 5. *Cont.*

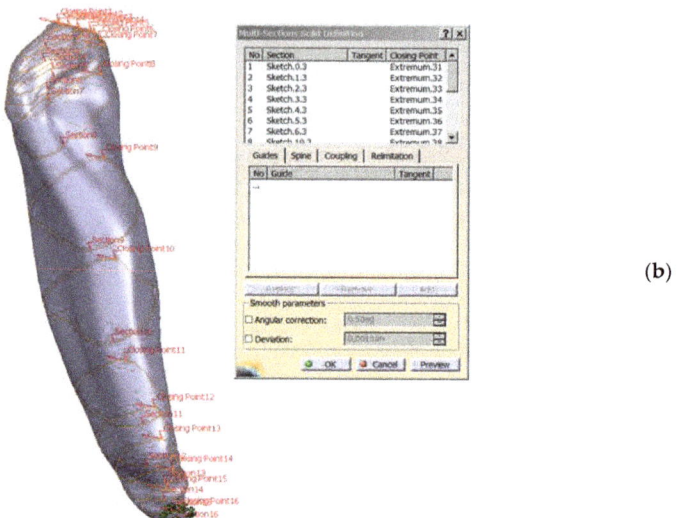

Figure 5. Construction of the canine's geometry: (**a**) necessary sections for the canine, based on CBCT; (**b**) the construction of the canine geometry based on the sections.

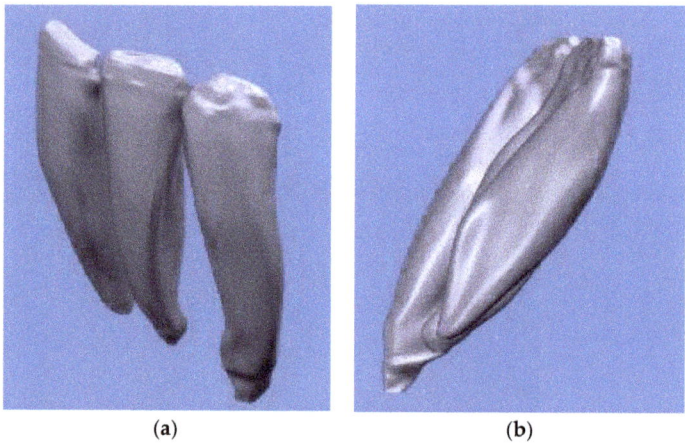

Figure 6. The model obtained for three teeth: (**a**) facial aspect (from left to right—the central incisor, the lateral incisor and the canine); (**b**) lingual aspect (from left to right—the central incisor, the lateral incisor and the canine).

Before performing the finite element analysis, the last element, the PDL's geometry, was created.

The periodontal ligament model was built based on 3D models of the mandible's periodontium and teeth, applying the principle that the ligament has two common contact surfaces, one with the tooth's radicular surface and the other with the mandibular bone, mediating the relation between the tooth and the bone [10]. The periodontal ligament is essential during orthodontic treatment for distributing tension and compression loading to the alveolar bone [11]. The modeling of the ligament was performed by calculating the average distance between the bone and the teeth on the CBCT images, and it resulted in a value of 0.4 mm. This thickness is comparable with the 0.4 mm thickness presented in previous studies in this field [12]. In our study, a maximum of 0.4 mm PDL thickness and a minimum of 0.05 mm PDL thickness were used to perform the analysis. Given that the CBCT images do not allow the use of a similar method for 3D modeling of the ligament, because the boundaries between bone, ligament and tooth are not clear, the construction of the ligament cannot be achieved by the same method of approximating the outline used for the bone and tooth. Thus, the average thickness was used as the main construction factor. A duplicate of the tooth was made, and the geometry was copied towards the exterior, at a distance equal to the thickness of the ligament. The tooth shape was subsequently extracted from the respective copy. In this way, the surface of the PDL from the tooth's side was obtained. Subsequently, the same duplicate intersected with the bone structure. In this way, we obtained a mandibular tooth socket of the mandibular bone.

C. Stages of periodontal ligament 3D modeling

1. In the first stage, the 3D model with the dimensions obtained after copying the contours from the CBCT was scaled down to the real dimensions. In these contours, each pixel in the CBCT image represents one millimeter in the Catia V5R20 program. The resolution of the introduced images was 1920 × 1080, so that each section was much more enlarged compared to reality. In order to convert the model to a realistic size, the real length was measured on the CBCT and then it was determined in the virtual model of the Catia V5R20 length. This resulted in a ratio of scaling factor $F_1 = 0.05309038$. This scaling factor was applied to the 3D model in order to ensure that the structure composed of the teeth, bone and PDL ligament reflected the real size. This factor was applied in turn to each component element (bone, teeth, periodontal ligament), and then the assembly was performed again in the *Assembly Design* module. The overall positioning was not altered because the relative positions to the coordinate system were also taken into account when resizing. After scaling, the appearance does not change, only the dimensions.

2. In the second stage, the external dimensions of the ligament were built based on the geometry of the tooth. This scaling allowed the use of the actual dimensions of the ligament and the next step was the modeling of the PDL. For the construction of the ligament, an artifice frequently used in 3D modeling was used. Starting from the shape of the tooth, scaling was applied to obtain a duplicate of the tooth (its shape) at a distance of 0.4 mm from the exterior. For this, the maximum diameter of the tooth was measured in different places. Subsequently, the average value was calculated and the thickness of the ligament was added twice, because it appears on both sides of the tooth. This represented the outer diameter of the ligament. Following this, the ratio between the tooth's diameter and the ligament's diameter was made. The value obtained, $F_2 = 1.023$, was used as a scaling factor to obtain the duplicate of the tooth, necessary for the construction of the ligament. In the Catia V5 R20 program, the *"scales"* function was used. Two parameters are needed for this function: the scale factor, 1.023, and the invariant point of scaling, meaning the scaling center. In our case, the central point of the tooth was chosen. It was built using a line that connected the tip with the apex of the tooth. On this line, the middle of the segment was highlighted and it was selected as an invariant point (reference) for scaling.

3. In the last stage, the tooth geometry was displaced from the obtained ligament, and later, the geometry of the ligament was displaced from the mandibular bone.

4. The ligaments for the 3 teeth were similar for the two incisors. The thickness of the ligament resulting from this procedure was not constant, but, on average, it was around 0.4 mm. We believe that this approach and construction of the tooth–ligament–bone assembly is the one that best approximates the biological nature of tissues. The obtained periodontal ligaments of the three teeth are illustrated in Figure 7.

Figure 7. Illustration of the 3 ligaments (from left to right—the central incisor, the lateral incisor and the canine).

5. After obtaining the geometric model of the assembly, we performed the finite element analysis to virtually test the response to various requests. The geometric model was discretized, and, with the help of the equations that describe the reunited phenomena in the form of a basic analytical model, the FEA was performed. The module used was Catia V5 R20—Generative Structural Analysis. The basic analytical model consists of a set of equations that describe the phenomenon studied and the behavior of the material under the action of external stresses. To these, boundary conditions were added. The boundary conditions describe the body's interaction with the environment. When the field sizes are variable over time, the analytical model must also include their initial conditions, meaning their condition at the beginning of the analysis. This analytical model is the basis for the development of the finite element analytical model, and its predictive variables are the basis of the numerical model's simulation performance. The analytical model must therefore capture this phenomenon of deformation under the action of external forces. It will contain, in this case, the definition relations of the normal unitary effort and the specific deformations. In order to be able to individualize the behavior of a certain material under the action of external loadings, a constitutive or material law must also be included in the analytical model. For us, this is Hooke's law, which shows that, in the case of an axially stressed solid material, as long as the external forces do not exceed a certain limit (flow limit), the unit forces are directly proportional to the specific deformations. Material properties were assigned to each component of the assembly in order to perform a finite element analysis.

Therefore, a specific material was used for each type of tissue. The value of E, the modulus of elasticity (Young's modulus), the breaking strength, the maximum deformation and the Poisson's ratio were all taken into consideration for our research. The modulus of elasticity is the characteristic of the material and is unique for each type of material (Table 1).

Table 1. Material properties.

Material	Young's Modulus	Poisson's Ratio
PDL	7.1×10^{-4} GPa [13]	0.4
Bone	140 GPa [14]	0.3
Tooth	20.3 GPa [14]	0.3

6. The operating rules for each element were established in the next step. Thus, a fixing constraint (clamp) was established for the fixed body, the mandibular bone. A clamp was used on the mesial side and one on the distal side. The tooth was considered mobile and the Generative Structural Analysis (GSA) module was used to apply a Fastened Connection constraint. For the PDL, a *Fastened Connection* constraint in the GSA module was also applied. The next step proceeded with the construction of the model for finite element analysis. This was done by applying boundary conditions and establishing fixed surfaces. For this, the *Analysis and Simulation/Generative Structural Analysis* workbench was opened. Through this module, the 3D model was further discretized and the conditions of demand and degrees of freedom were imposed. For the surfaces that were in contact, boundary conditions were imposed through the General Analysis Connection tool. This is a very powerful tool that can be used to connect any part of an assembly on an overall model, with or without a handling point. This tool allows any type of geometry to be connected, and it is a general way of connecting two components. The *Fastened Connection* rigid assembly connection tool was used to assign specific characteristics for simulation. Its purpose was to model a fixed connection between two bodies with a common boundary and the association of a geometric assembly constraint, or a General Analysis Connection type connection. Finally, there were 6 connections between the contact surfaces of the following components: central incisor and central incisor ligament; central incisor ligament and bone; lateral incisor and lateral incisor ligament; lateral incisor ligament and bone; canine and canine ligament; canine ligament and bone.

After the creation of these connections, the surfaces that were considered fixed were constrained with the *Clamp* tool. These constrained surfaces were the two lateral surfaces, the mesial and the distal surfaces of the mandibular bone. In establishing the values for the discretization network of previously built 3D models, the following parameters were used (Table 2):

1. Size—choosing or changing the size of finite elements;
2. Absolute Sag—the absolute value of the deviation from the boundary—meaning the maximum allowed deviation value for approximating the geometry of the model;
3. Element Type—the type of the finished element.

Table 2. Finite element characteristics for each component.

Element	Size (mm)	Absolute Sag (mm)	El. Type
Bone	0.8	0.5	Linear tetrahedral
PDL central incisor	0.3	0.1	Linear tetrahedral
Central incisor	1.272	0.203	Linear tetrahedral
PDL lateral incisor	0.3	0.1	Linear tetrahedral
Lateral incisor	1.313	0.21	Linear tetrahedral
PDL canine	0.3	0.1	Linear tetrahedral
Canine	1.484	0.237	Linear tetrahedral

A uniformly distributed force was applied to each tooth. The *Distributed force* tool was used. This force was applied to the midpoint of the buccal surface of the tooth crown, as shown in Figure 8.

The recommended position for applying the brackets was respected. The position of the forces for each tooth in part is represented in Figure 8 in the form of x.

Next, the results obtained from virtual experiments performed by simulation using the finite element method are presented.

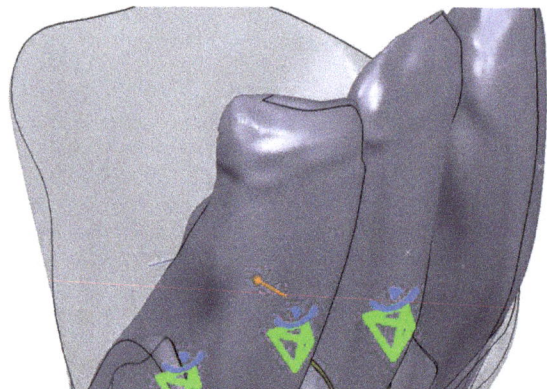

Figure 8. Place of application of the forces for each tooth, represented with x.

3. Results

The results present the outcome of the maximum stress in the bone, in the PDL and the absolute displacement after applying a force of 1 N and a force of 0.8 N in bodily direction on the central incisor, lateral incisor and canine. While applying different forces, the data from the same areas were recorded, in order to make a more accurate comparison. The sectioning plan of the tooth, its position, orientation and direction of sectioning were kept unchanged. The same tetrahedral finite elements were taken into account, in cases with no substantial changes in their geometry.

These forces (1 N and 0.8 N) were used due to the fact that they are very light forces that are preferably used on teeth with reduced periodontal support.

3.1. The Results after Applying the Forces on the Lateral Incisor

At the lateral incisive level, the forces in the lingual direction were applied in order to obtain a bodily movement. In Figure 9a,b, the maximum stresses in the bone and, respectively, in the PDL are represented. The color code applies to the whole ensemble.

As seen in Figure 9a, the highest stress in the bone and in the tooth was exhibited in area 5, the area with minimal ligament thickness. The application of a force in the lingual direction determined in the area with minimal ligament thickness the appearance of a contact point. This produced a lever in this type of movement and, clinically, the occurrence of root resorption could be seen. This explains the maximum stress from area 5, the increasing stress values in the areas with thinner PDL and the decrease in stress in area 6, the area where the PDL thickened again. Considering that area 7, the apex area, is a fixed point and area 1 is a mobile point, the stress should have decreased from area 1 to 7, but it increased in areas 5–6, the areas where the ligament was thinner.

Increasing the force from 0.8 N to 1 N resulted in a noticeable increase in the maximum stress in the PDL in area 1. The increase in the force by 25% resulted in a small increase in the stress (7.2%); therefore, it was not justified to use a force of 1 N instead of a force of 0.8 N. Analyzing the graphic representation, it was observed that the stress in the PDL decreased regularly, without variations, towards the apical area, contrary to the maximum stress identified in the bone (Figure 9b).

In area 5–6, the PDL was very thin, almost non-existent, and the displacement was accelerated by the decreased thickness of the PD. From area 1 to area 5, the appearance of a relatively horizontal curve confirmed that bodily displacement was obtained in this situation. Areas 6 and 7 exhibited lower displacement due to the fact that the apex was fixed to the periodontium. Otherwise, the two curves were similar: as the force increased, an increase in displacement was observed (Figure 9c). Therefore, when the force was increased from 0.8 to 1 N, the stress in the bone increased significantly compared to the values in the ligament; the two lines had a similar appearance for 0.8 N and 1 N. Therefore, stress levels increased in areas 4, 5 and 6, where the PDL was thinning, with a peak in zone 5, where the PDL was greatly diminished and it was not fully functional, and where the PDL was greatly diminished and it was partially functional. In area 7, its value returned to the tendency of the curve.

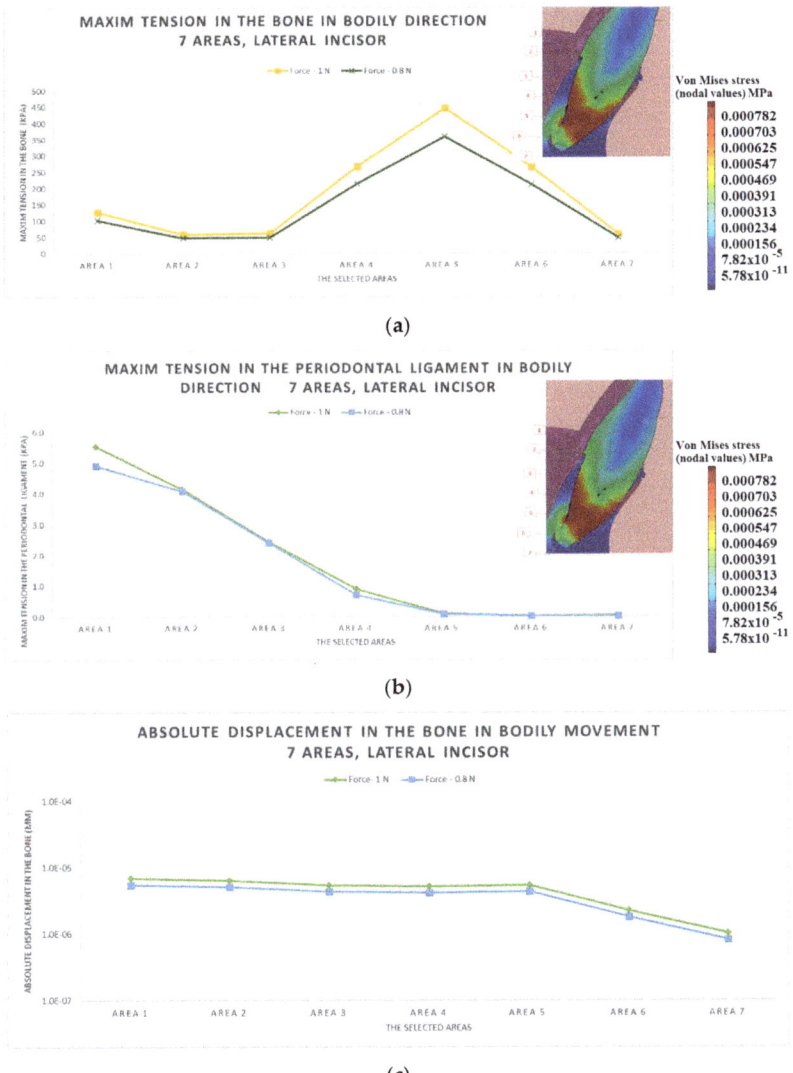

Figure 9. (**a**) Maximum stress in the bone for the lateral incisor. (**b**) Maximum stress in the PDL for the lateral incisor. (**c**) Absolute displacement in the bone in bodily movement for the lateral incisor.

3.2. The Results after Applying the Forces on the Central Incisor

The results respect the color code. The red color corresponds to the maximum stress identified in area 1. Analyzing the stress from one force to another (from 0.8 N to 1 N), as well as the differences for each zone, it was found that the highest stress was in area 1 and it decreased gradually towards area 7, the two curves being identical. Analyzing the stress distribution between area 1 and area 7, it was observed that the difference between the maximum stresses gradually decreased towards the apical area. In this area, the stress was roughly similar, and the differences were indistinguishable.

This phenomenon can be explained by the fact that the force is applied at the coronary level, and area 1 is closest to the force application area, while zone 7 is the furthest. Area 4 is the inflection area between the two curves, and areas 4–5 transition towards the apical area (Figure 10a).

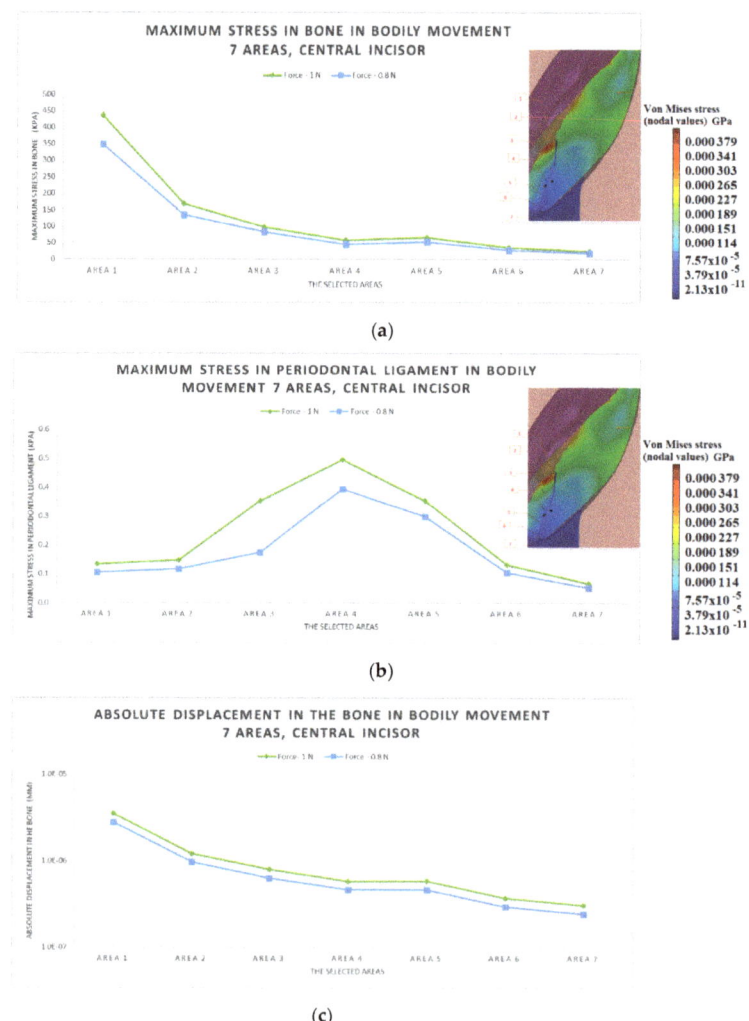

Figure 10. (**a**) Maximum stress in the bone in the central incisor. (**b**) Maximum stress in periodontal ligament in the central incisor. (**c**) Absolute displacement in the bone in bodily movement for the central incisor.

It was observed that stress depends on the PDL's thickness. Thus, the stress increased starting with area 3, reached its maximum in area 4 and decreased in zone 5. This phenomenon is explained by the fact that in area 4, there was a minimal ligament thickness; in areas 3 and 5, the PDL's thickness increased; and in area 6, the PDL displayed a relatively normal value. The two curves, both for the force of 1 N and for the force of 0.8 N, are similar in terms of shape; in addition, both exhibited increases in areas 3, 4 and 5. The maximum forces on the two areas of the tooth started approximately from the same area (Figure 10b).

Observing the amplitude of the displacement for each area, the greatest displacement was found in area 1 and the lowest in area 7. Thus, this was bodily displacement with tipping. These two curves had an identical appearance, without any change after an increase in force from 0.8 to 1 N (Figure 10c).

3.3. The Results after Applying the Forces on the Canine

Studying the appearance of the bodily movement in the case of two different forces (1 N and 0.8 N), it was found that the allure of the curve decreased from the cervical margin to the apical area. The maximum stress reached in area 4 is explained by the particular anatomic shape of this tooth (Figure 11a).

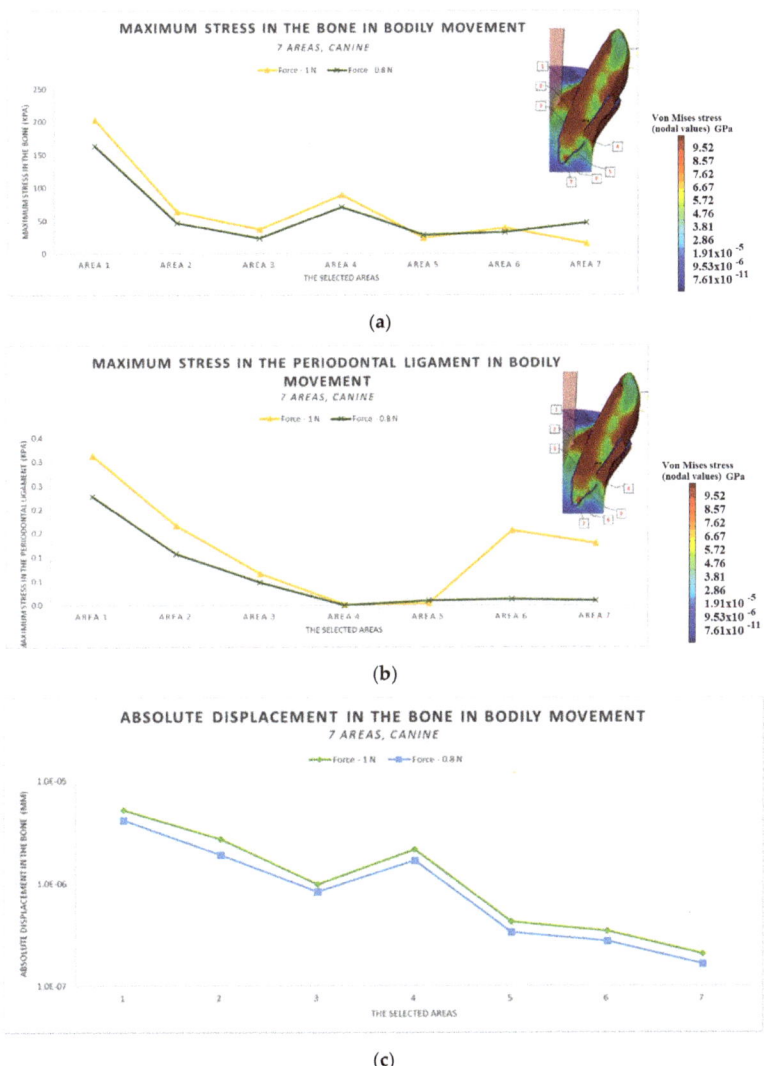

Figure 11. (a) Maximum stress in the bone for the canine. (b) Maximum stress in the periodontal ligament in the canine. (c) Absolute displacement in the bone in bodily movement.

The stress in the PDL for bodily movement decreases from the cervical margin towards the apical area. However, in the case of the 1 N force, a substantial increase appeared in areas 6 and 7, in the tooth's apical proximity, while in area 4 and 5, minimum values were registered (Figure 11b).

Regarding the bodily movement, a different type of displacement was observed in area 4. This phenomenon can be explained by the different anatomy of the tooth in area 4. The particular anatomy of the root influences the displacement, which normally should have been linear descendent (Figure 11c).

4. Discussion

The findings of this study suggest that the degree of displacement is correlated with the PDL's thickness. These results are in agreement with other studies [15].

Furthermore, stress increases in the PDL depending on the PDL's degree of thickness. In our study, for the maximum stress in the PDL, for the lateral incisor, it was observed that increasing the force by 25% determined an insignificant increase in the stress. Therefore, it was not justified to use a force of 1 N instead of a force of 0.8 N. These parameters depend on the particular anatomy that was captured by CBCT-based modeling and by object-oriented FEA (real case). These results are similar to those obtained in orthodontic practice.

From a medical point of view, it is recommended to use lower values because higher values could trigger root resorption, particularly in cases with abnormal root shape and increased root length [16,17]. Taking into consideration that the PDL has an absorption function and facilitates the displacement and alignment of teeth during orthodontic treatment [18] in the areas with decreased thickness, the displacement is accelerated or significantly increased. The correlation between the PDL's thickness and displacement is displayed throughout our study. Analyzing the three graphic representations for the lateral incisor (Figure 9a–c), it can be noticed that the displacement is higher in the area where the maximum bone stress is highest, and in area 5–6, the displacement is increased. This outcome is due to the minimal ligament thickness. Thus, these aspects should be well known before treatment planning in order to apply the necessary therapy according to the particularities of each patient. In order to analyze the effects of forces on the atypical anatomy of the root, FEA is a suitable, non-invasive method with results that can be subsequently applied in practice, in orthodontic treatment.

Regarding the type of displacement obtained, it can be observed, in the case of the three teeth (central incisor, lateral incisor and canine), that a bodily displacement with tipping was obtained. This aspect is highlighted by the shape of the curve. A horizontal line of the curve means that the movement is a pure bodily movement. Nevertheless, in our study, the line is descendent, which means that the movement is not a pure bodily movement, and it is a bodily displacement with tipping (Figures 9c, 10c and 11c).

Comparing the stress values in the ligament and in the bone, it was observed that the highest stress developed in the lateral incisor due to the minimal ligament thickness in area 4. Our results are comparable with the findings of Jing et al. [19], who found the highest stress in the central incisor. The results of our study would have been similar to those from previous studies, provided that the anatomy of the mandibular teeth would not be atypical compared to the maxillary teeth. This highlights the need for studies on the anatomical features of the teeth and periodontium, FEA being one of the most recommended methods. The highest maximum stress was found in area 1, in proximity to the cervical margin. This is in agreement with other studies [19].

The orthodontic–periodontal interrelationship is well known and systematically studied [20–22], particularly since the reduced periodontium is considered at risk regarding orthodontic treatment [23]; thus, the necessity of studies in this field is justified. This type of periodontium has low resistance to mechanical stress [24]. It is well known from practice and the literature that the most common root resorption occurs at the level of lateral and central incisors [17,25]. The explanation of this phenomenon is given by the radiological anatomy of these teeth (curved shape), the position on the arch (it is positioned with predilection towards lingual), the thinner cortex at this level and, especially in the case of the mandible, in the hypomochlion area, the PDL is thinner. Therefore, the forces will be concentrated at the center of rotation, which is dependent on the periodontal level. It is known that a reduced periodontium favors tipping movement and root resorption. A sufficient thickness of the PDL is also required to properly transmit the force through the bone and to correctly displace the teeth during orthodontic treatment. Studies show that the mandibular incisor region is the most common site and most prevalent for bone dehiscence and gingival recession [26,27], due to anatomical restrictions of the alveolar bone area, mainly in the antero–posterior aspect [28]. There has been discovered a correlation between alveolar bone loss and gingival recession after mandibular incisor retroclination [29].

Nevertheless, this study has some limitations. The model used for the FEA included only three teeth and their periodontium and they were considered isotropic. In addition, another limitation of the current study is that in orthodontics, the application of forces with a single point of application is very rare and can only be achieved with the use of specific techniques. Future research should consider analyzing the most widely used orthodontic techniques and extending the research on the entire mandible and even on the maxilla, in order to be able to compare the results. It should include several different characteristics for the PDL, such as an anisotropic ligament.

5. Conclusions

It was found that the maximum value of bone stress for the lateral incisor was in area 5, due to the minimal ligament thickness; it reached 448 KPa for the force of 1 N and 358 KPa for the force of 0.8 N. From a biomechanical point of view, a grade 2 lever was formed, due to the atypical root anatomy. The point of application of the resisting force (area 5, area with minimal PDL thickness) is between the fix point (tooth's root) and that of the active force (force applied to the tooth). Regarding the displacement of the lateral incisor, starting with area 5, the area with minimal ligament thickness, there was a decreasing tendency of the curve. In area 5, there was a risk of bone resorption occurrence due to increased stress, due to atypical root anatomy and longitudinal root depression. The increased displacement of the bone was determined by the minimal ligament thickness, as a result of the PDL's function, as the PDL optimizes the transfer of the forces and regulates orthodontic tooth movement. In the central incisor, the maximum bone stress was in area 1. In comparison, the maximum bone stress in the lateral incisor was in area 5, as a result of an atypical root anatomy. In areas 3, 4 and 5, areas with minimal ligament thickness, an increase in ligament tension was observed. Displacement was favored in areas 4–5 by the minimal ligament thickness; the curve had a constant horizontal appearance. This could determine root resorption. At the lateral incisor, the maximum stress in the bone in the area with the maximum increase for area 5 was 448 KPa, and in area 1, it was 128 KPa. In the canine, the maximum stress in the bone, in the area with the maximum increase, for area 4 was 89 KPa, and in area 1, it was 203 KPa. At the central incisor, the maximum stress in the bone, in the area with the maximum increase at the cervical margin for area 1 was 436 KPa.

Comparing the two curves of the central and lateral incisor, it was observed that:

1. The difference in increase is much higher at the lateral incisor (approx. 300 KPa), compared to the canine (approximately 100 KPa).
2. The difference in increase is explained by the specific mode of action of the force in the two cases. For the lateral incisor, the lever is of degree 2, with the force arm between zone 5 and zone 1; for the canine, the lever is of degree 1, and the arm of force between zone 1 and zone 4.
3. Due to the lever configuration, a compression force of the ligament appears in area 6–7 and this causes the PDL's increase.
4. In area 4, the risk of bone resorption due to increased stress is observed, as a result of atypical root anatomy. The minimal ligament thickness in area 4 causes increased displacement of the bone. This occurs as a result of the PDL's function, which optimizes the transfer of the forces and regulates orthodontic tooth movement.

5. The displacement is a bodily movement with tipping.

Considering that the stress is influenced by the tooth's morphology and the periodontal status, it can be concluded that the results obtained are similar to those from clinical practice. In addition, the outcome of this study demonstrates the correlation between the clinical results and FEA. Specific implications of orthodontic forces and biomechanical compression on the periodontal tissues are difficult to quantify; therefore, FEA is mandatory in order to achieve a treatment that is planned according to the patient's particular situation. This is even more important in patients with reduced periodontium that are predisposed to recessions.

Author Contributions: Conceptualization, I.-A.S. and M.A.; methodology, S.M.S. and M.A.; software, D.-C.K.-N.; validation, S.M.S. and N.C.; formal analysis, I.-A.S.; investigation, S.T. and A.M.; resources, S.M., M.P. and N.C.; data curation, I.-A.S. and A.M.; writing—original draft preparation, I.-A.S. and M.A.; writing—review and editing, I.-A.S. and N.C.; visualization, A.M., I.L. and M.A.; supervision, S.M.S., M.A., N.C. and I.L.; project administration, M.P. and S.M.; funding acquisition, S.M. All authors have read and agreed to the published version of the manuscript.

Funding: This study was partially funded through a PhD scholarship offered by the Romanian Ministry of Education to Ioana-Andreea Sioustis, a PhD student at the University of Medicine and Pharmacy "Grigore T. Popa" Iasi, Romania and partially by Romanian Ministry of Education and Research, CNCS-UEFISCDI, project number PN-III-P1-1.1-TE-2019-1921, within PNCDI III.

Institutional Review Board Statement: The study was conducted according to the guidelines of the Declaration of Helsinki and approved by the Ethics Committee of the University of Medicine and Pharmacy from Iasi, Romania (Protocol identification code 29.01.2020/2540).

Informed Consent Statement: Any research article describing a study involving humans should contain this statement. Please add "Informed consent was obtained from all subjects involved in the study." OR "Patient consent was waived due to REASON (please provide a detailed justification)." OR "Not applicable." for studies not involving humans. You might also choose to exclude this statement if the study did not involve humans.

Data Availability Statement: The data used to support the findings of this study are available from the correspondence authors upon request.

Acknowledgments: Sorina Mihaela Solomon and Ionut Luchian have a contribution equal to that of the first author.

Conflicts of Interest: The authors declare no conflict of interest.

References

1. Bramanti, E.; Cervino, G.; Lauritano, F.; Fiorillo, L.; D'Amico, C.; Sambataro, S.; Denaro, D.; Famà, F.; Ierardo, G.; Polimeni, A.; et al. FEM and Von Mises Analysis on Prosthetic Crowns Structural Elements: Evaluation of Different Applied Materials. *Sci. World J.* **2017**, *3*. [CrossRef]
2. Lauritano, F.; Runci, M.; Cervino, G.; Fiorillo, L.; Bramanti, E.; Cicciù, M. Three-dimensional evaluation of different prosthesis retention systems using finite element analysis and the Von Mises stress test. *Minerva Stomatol.* **2016**, *65*, 353–367.
3. Goel, V.K.; Khera, S.C.; Ralson, J.L.; Chang, K.H. Stresses at the dentino-enamel junction of human teeth: A finite element investigation. *J. Prosthet. Dent.* **1991**, *66*, 451–459. [CrossRef]
4. Butnaru-Moldoveanu, S.A.; Munteanu, F.; Forna, N.C. Virtual Bone Augmentation in Atrophic Mandible to Assess Optimal Implant-Prosthetic Rehabilitation—A Finite Element Study. *Appl. Sci.* **2020**, *10*, 401. [CrossRef]
5. Carpegna, G.; Alovisi, M.; Paolino, D.S.; Marchetti, A.; Gibello, U.; Scotti, N.; Pasqualini, D.; Scattina, A.; Chiandussi, G.; Berutti, E. Evaluation of Pressure Distribution against Root Canal Walls of NiTi Rotary Instruments by Finite Element Analysis. *Appl. Sci.* **2020**, *10*, 2981. [CrossRef]
6. Huang, H.-L.; Tsai, M.-T.; Yang, S.-G.; Su, K.-C.; Shen, Y.-W.; Hsu, J.-T. Mandible Integrity and Material Properties of the Periodontal Ligament during Orthodontic Tooth Movement: A Finite-Element Study. *Appl. Sci.* **2020**, *10*, 2980. [CrossRef]
7. Luchian, I.; Moscalu, M.; Martu, I.; Curca, R.; Vata, I.; Stirbu, C.; Tatarciuc, M.; Martu, S. A FEM Study regarding the Predictability of Molar Uprighting Associated with Periodontal Disease. *Int. J. Med. Dent.* **2018**, *22*, 183–188.
8. Krishnan, V.; Davidovitch, Z. Cellular, molecular, and tissue-level reactions to orthodontic force. *Am. J. Orthod. Dentofac. Orthop.* **2006**, *129*, 469.e1–469.e32. [CrossRef] [PubMed]

9. Theerasopon, P.; Kosuwan, W.; Charoemratrote, C. Stress assessment of mandibular incisor intrusion during initial leveling in continuous arch system with different archwire shapes of superelastic nickel-titanium: A three-dimensional finite element study. *Int. J. Health Allied Sci.* **2019**, *8*, 92–97.
10. Nishihira, M.; Yamamoto, K.; Sato, Y.; Ishikawa, H.; Natali, A.N. Mechanics of periodontal ligament. *Dental Biomech.* **2003**, 20–34. [CrossRef]
11. Dorow, C.; Krstin, N.; Sander, F.G. Experiments to determine the material properties of the periodontal ligament. *J. Orofac. Orthop.* **2002**, *63*, 94–104. [CrossRef] [PubMed]
12. Schmidt, F.; Lapatki, B.G. Effect of variable periodontal ligament thickness and its non-linear material properties on the location of a tooth's centre of resistance. *J. Biomech.* **2019**, *94*, 211–218. [CrossRef]
13. Xia, Z.; Jiang, F.; Chen, J. Estimation of periodontal ligament's equivalent mechanical parameters for finite element modeling. *Am. J. Orthod. Dentofac. Orthop.* **2013**, *143*, 486–491. [CrossRef] [PubMed]
14. Singh, J.R.; Kambalyal, P.; Jain, M.; Khandelwal, P. Revolution in Orthodontics: Finite element analysis. *J. Int. Soc. Prev. Community Dent.* **2016**, *6*, 110–114. [CrossRef]
15. Li, Y.; Jacox, L.A.; Little, S.H.; Ko, C.C. Orthodontic tooth movement: The biology and clinical implications. *Kaohsiung J. Med. Sci.* **2018**, *34*, 207–214. [CrossRef] [PubMed]
16. Sameshima, G.T.; Sinclair, P.M. Predicting and preventing root resorption: Part I. Diagnostic factors. *Am. J. Orthod. Dentofac.* **2001**, *119*, 505–510. [CrossRef]
17. Jung, Y.H.; Cho, B.H. External root resorption after orthodontic treatment: A study of contributing factors. *Imaging Sci. Dent.* **2011**, *41*, 17–21. [CrossRef]
18. Nakamura, N.; Ito, A.; Kimura, T.; Kishida, A. Extracellular Matrix Induces Periodontal Ligament Reconstruction In Vivo. *Int. J. Mol. Sci.* **2019**, *20*, 3277. [CrossRef]
19. Jing, Y.; Han, X.; Cheng, B.; Bai, D. Three-dimensional FEM analysis of stress distribution in dynamic maxillary canine movement. *Sci. Bull.* **2013**, *58*, 2454–2459. [CrossRef]
20. Calniceanu, H.; Stratul, S.; Rusu, D.; Jianu, A.; Boariu, M.; Nica, L.; Ogodescu, A.; Sima, L.; Bolintineanu, S.; Anghel, A.; et al. Changes in clinical and microbiological parameters of the periodontium during initial stages of orthodontic movement in patients with treated severe periodontitis: A longitudinal site-level analysis. *Exp. Ther. Med.* **2020**, *6*, 199. [CrossRef]
21. Machoy, M.; Szyszka-Sommerfeld, L.; Koprowski, R.; Wawrzyk, A.; Woźniak, K.; Wilczyński, S. Assessment of Periodontium Temperature Changes under Orthodontic Force by Using Objective and Automatic Classifier. *Appl. Sci.* **2021**, *11*, 2634. [CrossRef]
22. Sioustis, I.-A.; Martu, M.-A.; Aminov, L.; Pavel, M.; Cianga, P.; Kappenberg-Nitescu, D.C.; Luchian, I.; Solomon, S.M.; Martu, S. Salivary Metalloproteinase-8 and Metalloproteinase-9 Evaluation in Patients Undergoing Fixed Orthodontic Treatment before and after Periodontal Therapy. *Int. J. Environ. Res. Public Health* **2021**, *18*, 1583. [CrossRef]
23. Rasperini, G.; Acunzo, R.; Cannalire, P.; Farronato, G. Influence of Periodontal Biotype on Root Surface Exposure During Orthodontic Treatment: A Preliminary Study. *Int. J. Periodontics Restor. Dent.* **2015**, *35*, 655–675. [CrossRef]
24. Gorbunkova, A.; Pagni, G.; Brizhak, A.; Farronato, G.; Rasperini, G. Impact of Orthodontic Treatment on Periodontal Tissues: A Narrative Review of Multidisciplinary Literature. *Int. J. Dent.* **2016**, *4723589*. [CrossRef] [PubMed]
25. Abuabara, A. Biomechanical aspects of external root resorption in orthodontic therapy. *Med. Oral Patol. Oral Cirugía Bucal* **2007**, *12*, 610–613.
26. Albandar, J.M. Global risk factors and risk indicators for periodontal diseases. *Periodontology 2000* **2002**, *29*, 177–206. [CrossRef]
27. Wehrbein, H.; Bauer, W.; Diedrich, P. Mandibular incisors, alveolar bone and symphysis after orthodontic treatment: A retrospective study. *Am. J. Orthod. Dentofac. Orthop.* **1996**, *110*, 239–246. [CrossRef]
28. Swasty, D.; Lee, J.S.; Huang, J.C.; Maki, K.; Gansky, S.A.; Hatcher, D.; Miller, A.J. Anthropometric analysis of the human mandibular cortical bone as assessed by cone-beam computed tomography. *J. Oral Maxillofac. Surg.* **2009**, *67*, 491–500. [CrossRef]
29. Vasconcelos, G.; Kjellsen, K.; Preus, H.; Vandevska-Radunovic, V.; Hansen, B.F. Prevalence and severity of vestibular recession in mandibular incisors after orthodontic treatment: A case-control retrospective study. *Angle Orthod.* **2012**, *82*, 42–47. [CrossRef]

Article

Inlay-Retained Dental Bridges—A Finite Element Analysis

Monica Tatarciuc [1], George Alexandru Maftei [2,*], Anca Vitalariu [1,*], Ionut Luchian [3,†], Ioana Martu [1,†] and Diana Diaconu-Popa [1]

1. Department of Dental Technology, "Grigore T. Popa" University of Medicine and Pharmacy, 16 Universitatii Str., 700115 Iasi, Romania; monica.tatarciuc@umfiasi.ro (M.T.); ioana.martu@umfiasi.ro (I.M.); antonela.diaconu@umfiasi.ro (D.D.-P.)
2. Department of Oral Medicine, "Grigore T. Popa" University of Medicine and Pharmacy, 16 Universitatii Str., 700115 Iasi, Romania
3. Department of Periodontology, "Grigore T. Popa" University of Medicine and Pharmacy, 16 Universitatii Str., 700115 Iasi, Romania; ionut.luchian@umfiasi.ro
* Correspondence: george.maftei.gm@gmail.com (G.A.M.); anca.vitalariu@umfiasi.ro (A.V.); Tel.: +40-721414248 (G.A.M.); +40-232301618 (A.V.)
† Authors with equal contribution as the first author.

Abstract: Inlay-retained dental bridges can be a viable minimally invasive alternative when patients reject the idea of implant therapy or conventional retained full-coverage fixed dental prostheses, which require more tooth preparation. Inlay-retained dental bridges are indicated in patients with good oral hygiene, low susceptibility to caries, and a minimum coronal tooth height of 5 mm. The present study aims to evaluate, through the finite element method (FEM), the stability of these types of dental bridges and the stresses on the supporting teeth, under the action of masticatory forces. The analysis revealed the distribution of the load on the bridge elements and on the retainers, highlighting the areas of maximum pressure. The results of our study demonstrate that the stress determined by the loading force cannot cause damage to the prosthetic device or to abutment teeth. Thus, it can be considered an optimal economical solution for treating class III Kennedy edentation in young patients or as a provisional pre-implant rehabilitation option. However, special attention must be paid to its design, especially in the connection area between the bridge elements, because the connectors and the retainers represent the weakest parts.

Keywords: dental bridges; inlays; finite element analysis

1. Introduction

A dental prosthesis is an artificial substitute, which has the role of reestablishing the dental arch interrupted by edentation and restoring the functions of the dento-maxillary system. Furthermore, it is a biological infrastructure with the role of support and aggregation for prosthetic construction. Therefore, when analyzing the longevity of a prosthetic appliance, we must follow the two structural elements equally, which respond differently to demands and have different behaviors over time [1].

The treatment solutions for a single-tooth edentation are multiple and can be chosen according to several criteria: esthetics, mechanical resistance, the degree of teeth damage and, last, but not least, the patients' wishes. Possible treatment options in this situation can be metal–ceramic, all-ceramic, direct or indirect fiber-reinforced composite fixed dental prostheses, minimally invasive dental bridges or implants [2].

Posterior fixed partial dentures have different biomechanical behavior according to the restorative materials used. For example, some authors report that acrylic resin fixed prosthetic restorations can lessen the stress level in the connector region, and resin composite dental bridges can diminish the magnitude of stress on the layer of cement [3].

Inlay-retained fixed dental prostheses may indicate when adjacent teeth have been previously restored and when implant placement is not possible or not indicated. Inlay-retained bridges are also a good option in patients with good oral hygiene and low susceptibility to caries, with a minimum coronal tooth height of 5 mm, parallel abutments, and a maximum mesiodistal edentulous space of 12 mm. Contraindications include severe dental malpositions, the absence of enamel on the preparation margins, extensive crown defects and mobility of abutment-teeth [4,5].

A particularly important aspect that influences the clinical longevity of the minimally invasive bridges is mechanical resistance. This parameter depends on the dental materials used, design, and developing tensions, both at the bridge elements and at the abutment teeth [6].

Finite element analysis is a method that has as a main objective the modeling and description of mechanical behavior of elements with complex geometry, with the added advantage of the simplicity of basic concepts; the mathematical model thus realized includes certain working hypotheses, simplifications and generalizations [7]. To evaluate the stability of dental bridges and the stress they exert on the abutment teeth under conditions of variable masticatory demands, a working hypothesis can be simulated in the situation of a reduced partial edentation.

Inlay retained bridges are highly appreciated in the prosthodontic clinical practice since they are minimally invasive for the biological support tissues. However, from a biomechanical perspective, it is not clear whether they are as reliable as other therapeutic options [5,6]. Moreover, it is not clear whether the stress determined by the loading force can cause damage to the prosthetic device or to the abutment teeth, thus influencing the longevity of this restoration.

Because testing biomechanical parameters is not possible in the oral cavity without affecting the integrity of hard and soft tissues, the finite element method (FEM) is a useful way to appreciate the strengths and weaknesses of a prosthetic device and the stress induced into the dental support in various circumstances [8].

Considering all this and the multitude of design possibilities of the prosthetic appliance and possible materials available for the restoration, further studies are needed to clarify the indications and limitations of fixed dental bridges to maximize treatment outcome.

The present study aims to analyze stress levels at a three-element dental bridge, with inlays used as retainers to evaluate tension levels on abutment teeth and analyze the maximum tensions applied to abutment teeth and the dental prosthesis, considering the dynamic action of masticatory forces.

For this purpose, we used finite element analysis, a method that provides an image of the distribution of forces, both at the artificial substitute and the biological support level [9].

2. Materials and Methods

The finite element method attempts to approximate a solution to a problem by admitting that the domain is divided into subdomains or finite elements, with simple geometric shapes and the function of an unknown state variable that is defined around each element. The operation of choosing the number and type of finite elements, corroborated with the division of the domain into several finite elements, is called discretization. When preparing the analysis model, one must take into account that the shape and dimensions of the design influence the accuracy and time of analysis; in this sense, for a given problem, there are several variants of analysis models.

Description of the geometric model and finite element modeling was first performed, a stage that includes: modeling of material characteristics, choosing the finite elements and introducing properties, generating the finite element structure, introduction of limit conditions and forces. Analysis and solution of the finite element model involved setting the solving parameters and then visualizing the states and variations of the parameters.

Images of molars, canines, premolars and incisors were taken as a reference; their dimensions were made on a scale, according to the dimensions of the molars on computed tomography (CT) images. Based on the mandibular model, the absence of tooth 3.5 was simulated. To simplify the finite element analysis, only the dental elements 3.3, 3.4, 3.6 and 3.7 were preserved from the assembly.

For the application of optimal forces to simulate physiological masticatory load, the main muscles of the mastication process, the masseter muscles and lateral and medial pterygoid muscles were taken into account (Figure 1).

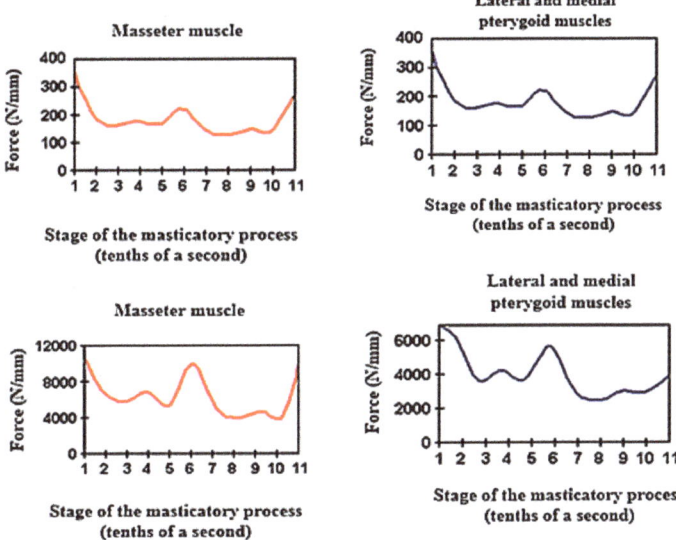

Figure 1. Variation of forces and moments developed by the masseter pterygoid muscles during the masticatory process (a mastication cycle).

The average values of masticatory forces vary from individual to individual and are also influenced by food consistency. In this study, we applied an average force value of 220 N in the premolar area and 400 N in the molar area, introduced in the Autodesk Simulation Mechanical 2014 program (2014, San Rafael, California, US). The model was subjected to loading with a vertical force, applied, in turn, on the occlusal surfaces of each element of the dental bridge: both retainers and pontic. The modeling of an inlay-retained dental bridge was performed, with retainers on teeth 3.4 and 3.6 and pontic on 3.5 (Figure 2). The material used for the bridge was a titanium alloy (Ti6Al4V) (Dentaurum, Germany).

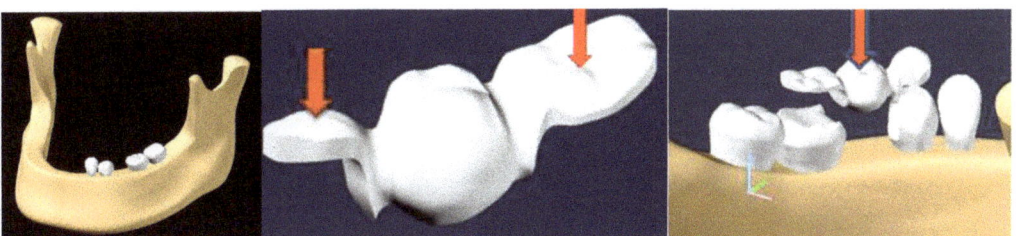

Figure 2. The models of the mandible and the bridge.

The force was applied first in the contact area of the occlusal surface of the inlay at the level of 3.4, then at the occlusal face of 3.6 and then on 3.5, occlusal surface, simulating the pressure induced by the food fragment during the masticatory act.

FEM analysis consists of a mesh realization by splitting a solid volume into finite elements of parallelepiped or tetrahedron shape. Each element behaves individually, with the same characteristics as the base material. Depending on the pressure applied to every element, a specific load or temperature will be exerted and transmitted to the adjacent elements through nodes. For an enhanced accuracy of results, a condition was imposed for the mesh realization: the length between two nodes must always be the same. The model was exported to Autodesk Simulation Mechanical 2014 with a file ending in *. * Sat. These files are opened one by one in Autodesk Simulation Mechanical 2014. After determining the type of analysis (static stress), the mesh command was given.

Depending on the objectives of the analysis, the used material properties were: modulus of elasticity, Poisson's ratio and density. The values for these parameters were taken from the literature [9] (Table 1).

Table 1. Material characteristics for each element of the structure subjected to finite element analysis.

Element	Modulus of Elasticity (MPa)	Poisson's Ratio	Density (kg/m^3)
Bone	13,800	0.30	1450
Dentine	18,600	0.31	1900
Ti6Al4V alloy	110,000	0.40	4381

We analyzed the tensions induced as a result of the application of these forces, on each element of the bridge, in the contact areas located between retainers and pontic and also on abutment teeth.

For the area where the bridge is in contact with the food fragment, during mastication, it is considered that there is no degree of freedom, and restrictions will be placed for all six movements. In Autodesk Simulation Mechanical 2014, these types of supports are represented by triangles.

3. Results

Once all the input data of the models were established, the analyze command was given. The results of the finite element analysis follow, according to the determination of loads applied to each area of the dental bridge. First, the distribution of deviatoric stresses (von Mises tensions) at the pontic level was analyzed when the reaction force, which opposes the muscular force, is located in the area of the missing premolar, 3.5 (Figure 3).

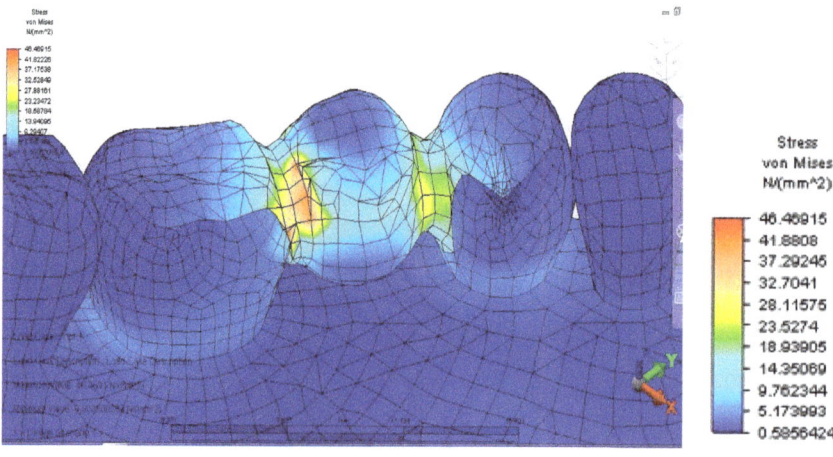

Figure 3. Tensions distribution at the bridge level, when the reaction force is placed in the 3.5 area.

From the model, we observed a higher tension on the prosthetic construction at the margin between the bridge and abutments. This phenomenon is related to the difference between the higher modulus of elasticity of the bridge alloy and the modulus of elasticity of abutment teeth. Furthermore, higher tensions develop at the junction between the bridge elements, and the maximum value is 46,469 MPa, between molar 3.6 and premolar 3.5. This high value of stress is explained by the smaller surface of the junction area.

When analyzing stress values on abutment teeth, respectively molar 3.6 and premolar 3.4, they have a maximum value in the cervical area of the junction between teeth and bridge: the value of 14.456 MPa is equally distributed on the two abutment teeth (Figures 4 and 5).

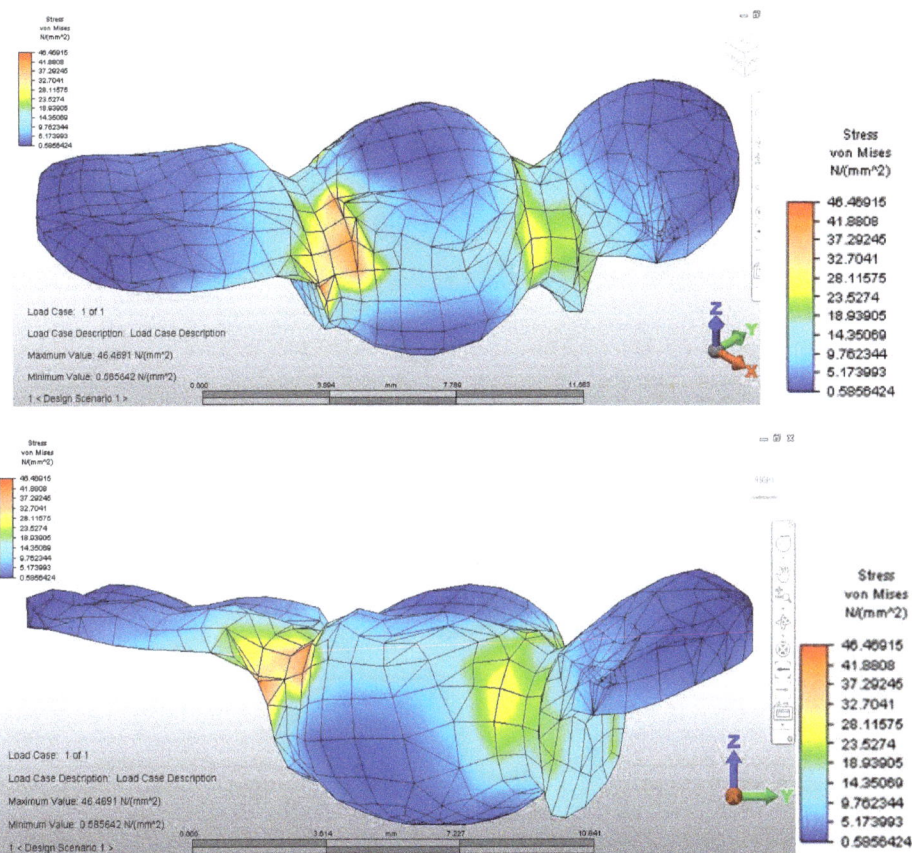

Figure 4. *Cont.*

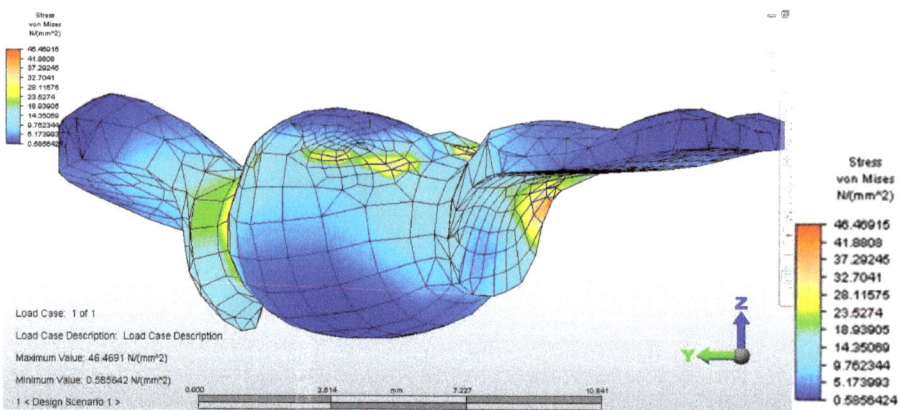

Figure 4. Tension distribution at the bridge level in three directions: axonometric, right and left, when the force is placed in the 3.5 area.

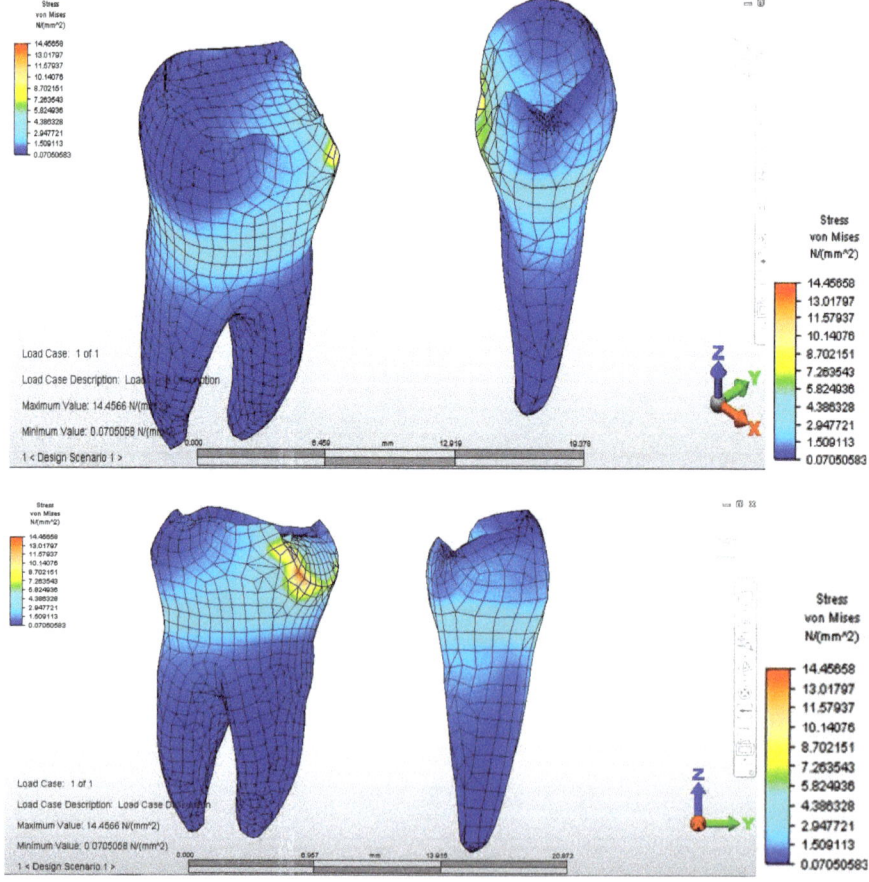

Figure 5. *Cont.*

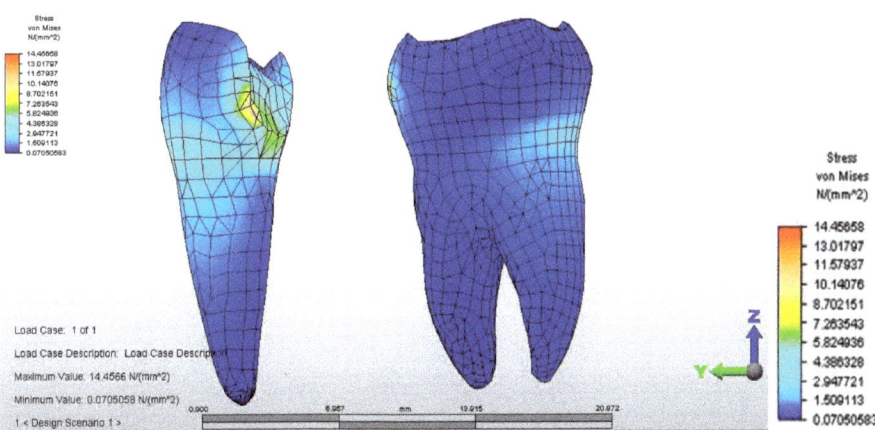

Figure 5. Tension distribution at the abutments level in three directions: axonometric, right and left when the force is placed in the 3.5 area.

Distribution of tensions at bridge level was then observed in the situation where the reaction force, which opposes the muscular force, is placed in the area of molar 3.6 (Figure 6).

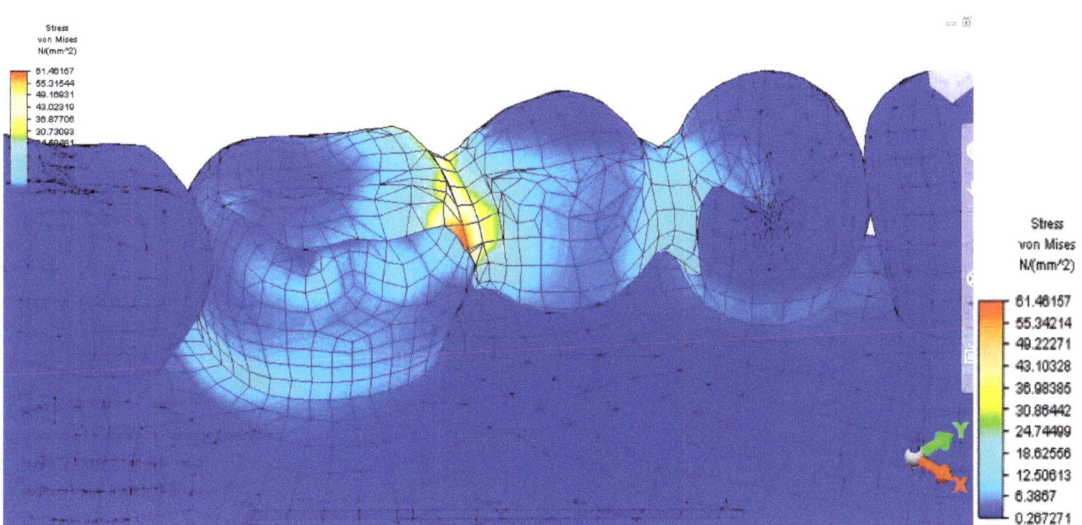

Figure 6. Tension distribution at the bridge level when the reaction force is placed in the 3.6 area.

This situation yielded a higher tension observed at the level of the bridge and at the mandibular bone level: the load is higher in the area of molar 3.6, where the reaction force is also applied.

The distribution of stresses at the level of the dental crown revealed that there is increased stress at the junction area between the pontic and the retainer, on molar 3.6; its value of 61.461 MPa, being 1.5 times higher than in the previous case. Abutment teeth are subjected to a maximum tension of 15.054 MPa, the higher value being encountered in molar 3.6, both in the contact area with the bridge and in the area of contact with the mandibular bone, toward molar 3.7 (Figures 7 and 8).

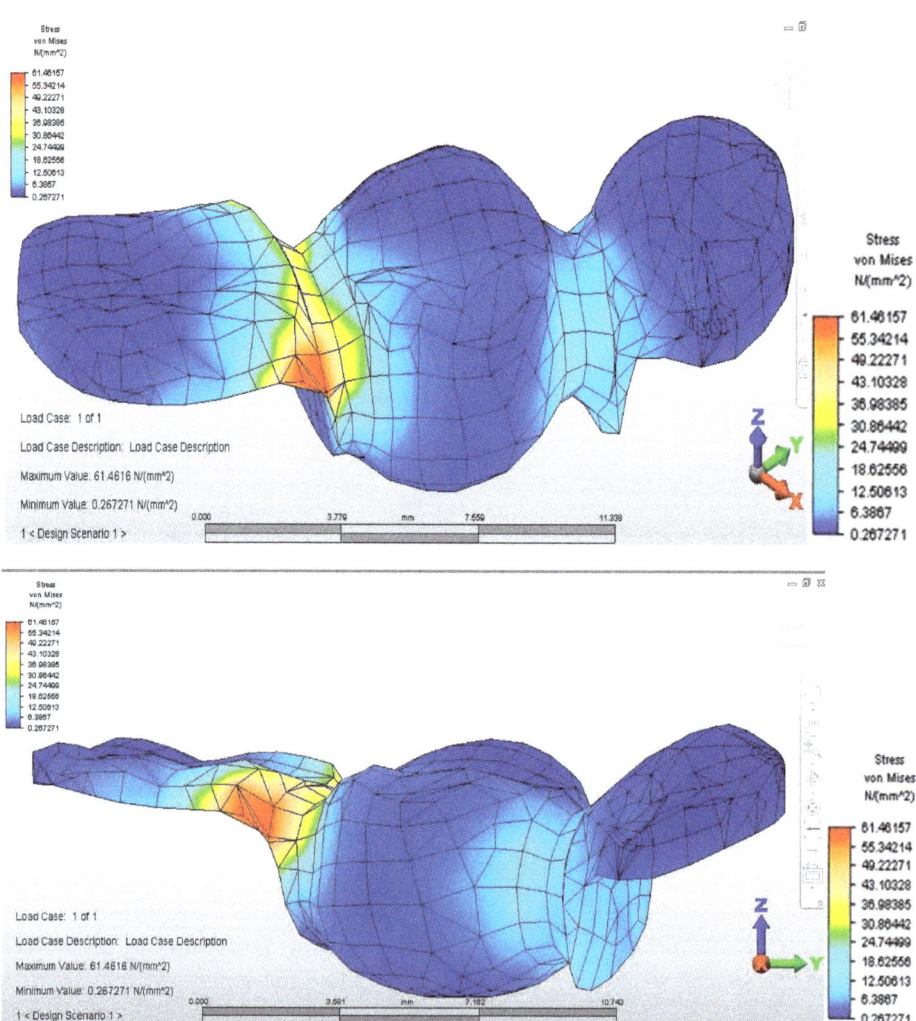

Figure 7. *Cont.*

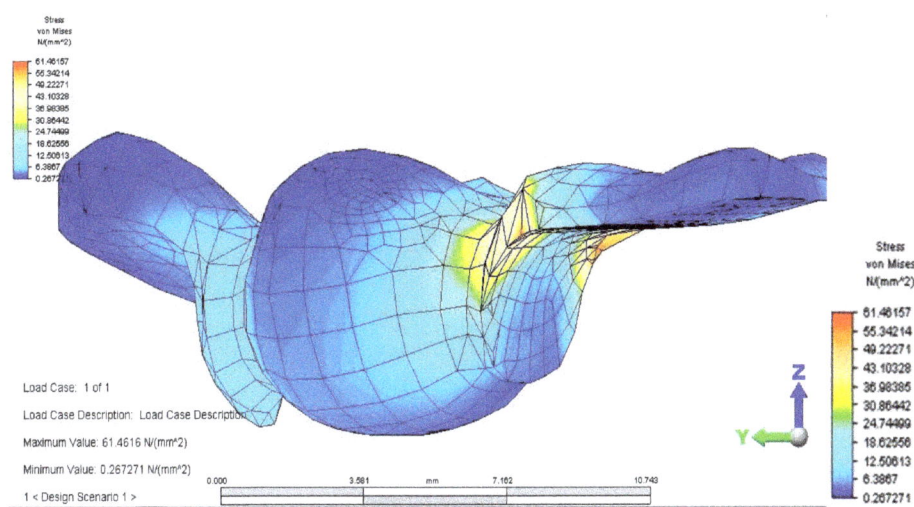

Figure 7. Tension distribution at the bridge level in three directions: axonometric, right and left, when the force is placed in the 3.6 area.

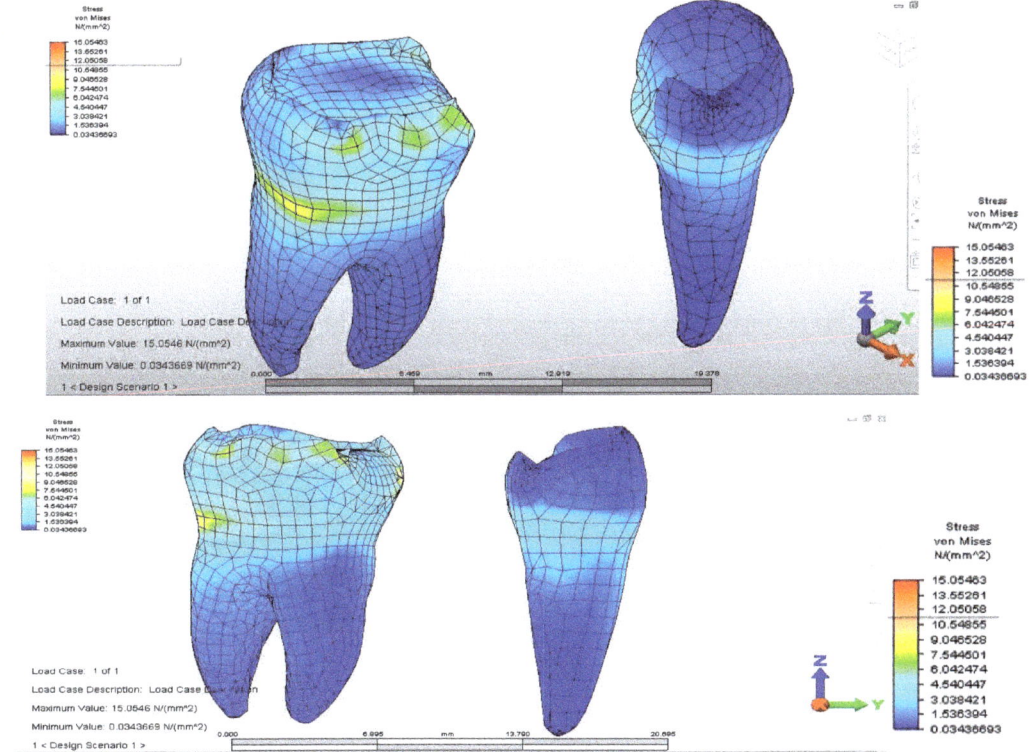

Figure 8. *Cont.*

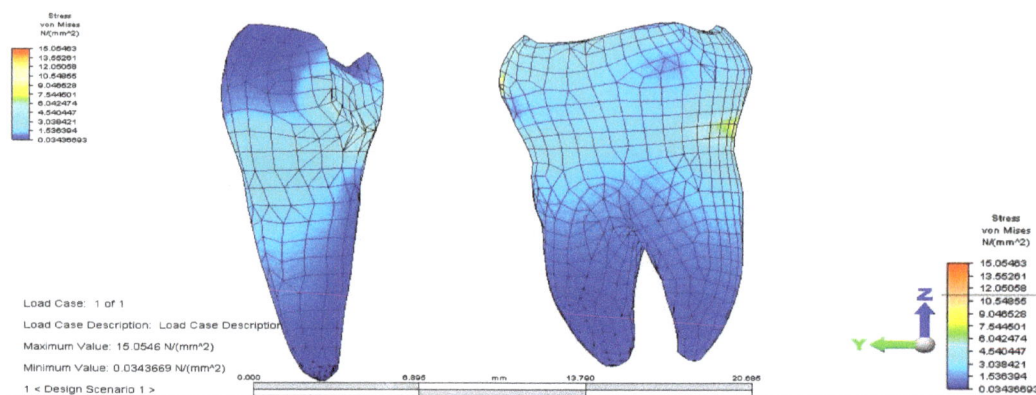

Figure 8. Tension distribution at abutments level in three directions: axonometric, right and left when the force is placed in the 3.6 area.

Regarding analyzing tensions in the mandible-abutments-dental bridge complex, when the reaction force acts on premolar 3.4, a tension jump is observed between the dental bridge and the teeth. This inequality is more evident at the junction between premolar 3.5 and the inlay applied on the occlusal surface of premolar 3.4. As a result of the reaction force exerted on tooth 3.4, the mandibular bone also supports a significantly higher tension (Figure 9).

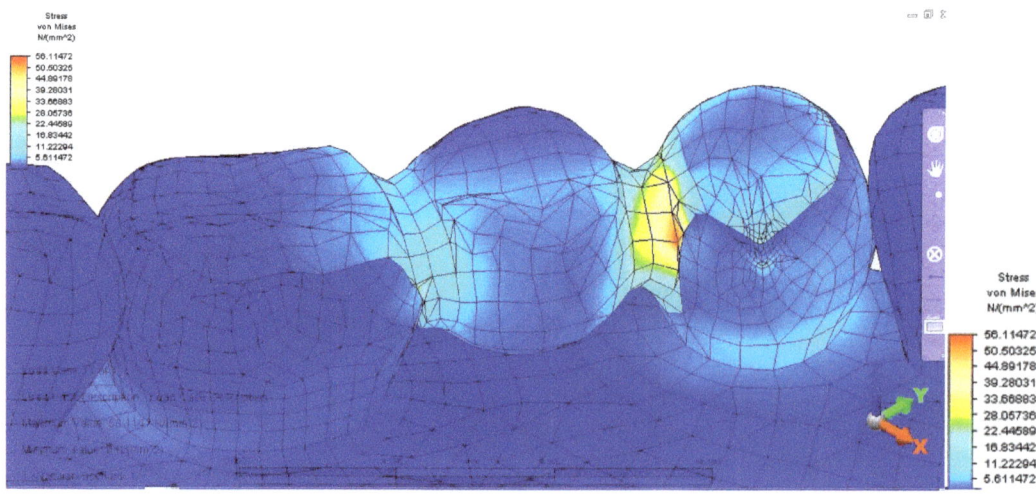

Figure 9. Tension distribution at the bridge when the reaction force is placed in the 3.4 area.

The dental bridge has an asymmetric stress distribution due to the reaction force, which has a maximum value of 56.114 MPa. However, the value is lower when the load is applied on the molar but higher when the force acts on the pontic. This is due to the greater distance of the premolar from the rotation center and also as a result of the direct action of the reaction force on premolar 3.4.

The pressure on abutments is also unevenly distributed. Stress distribution has a maximum value (12.553 MPa) on the premolar on which the reaction force acts. In addition, as in the case of applying for support on the molar, a higher value of tension is observed in the upper area of the root of premolar 3.4, towards the canine.

Both in this situation and in the previous case, the direct action of the reaction force on abutments also determines a tilt of the supporting teeth (Figures 10 and 11).

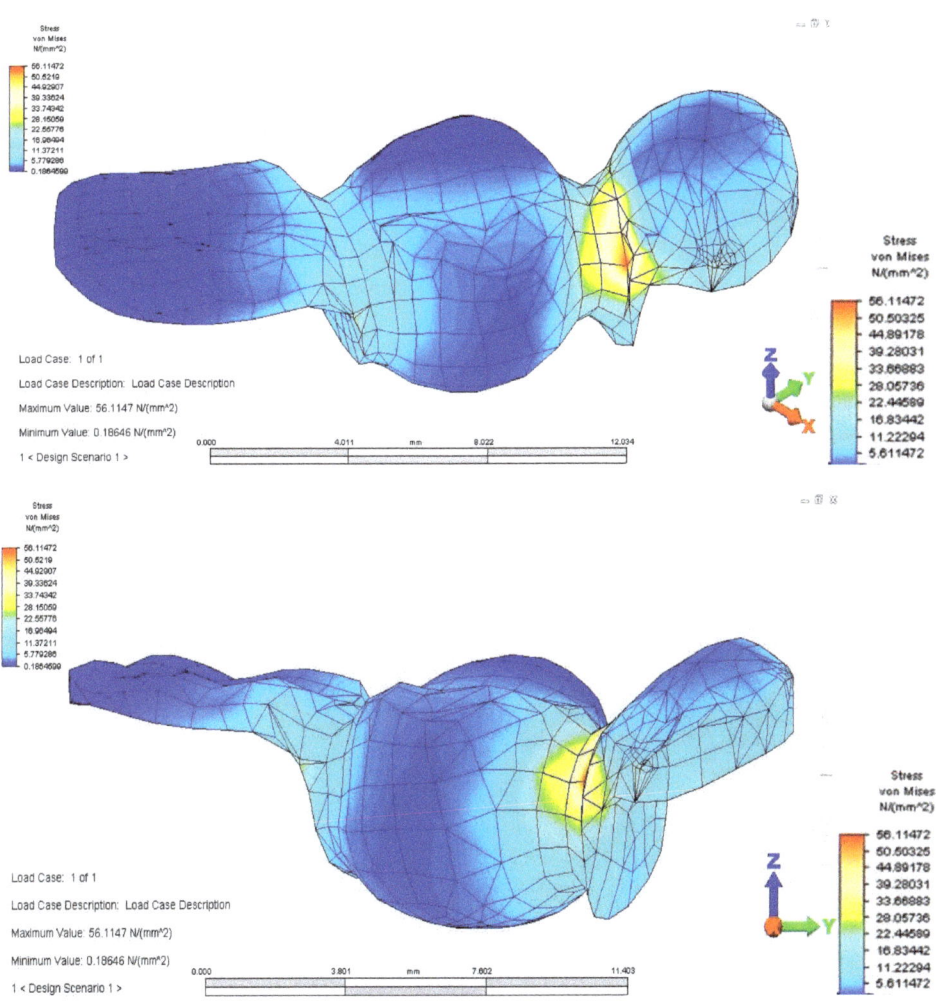

Figure 10. *Cont.*

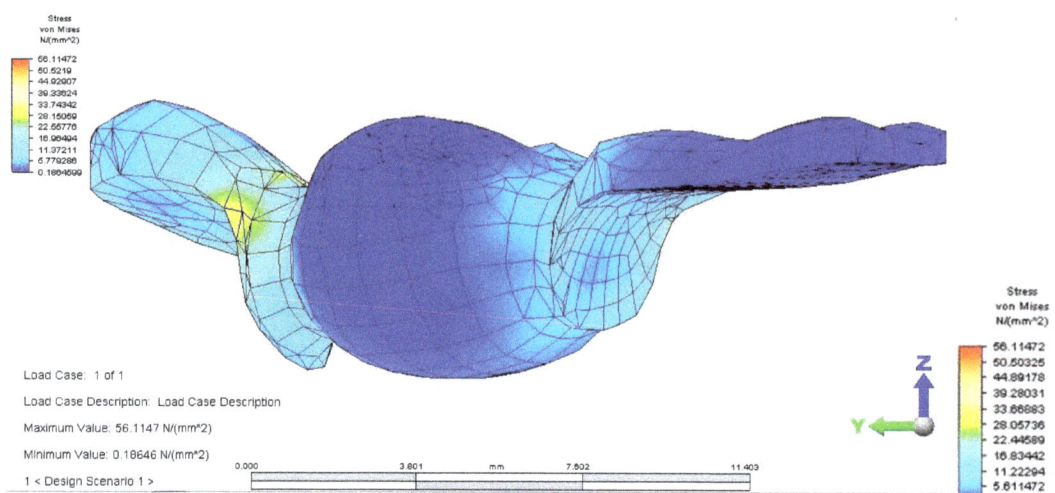

Figure 10. Tension distribution at the bridge level in three directions: axonometric, right and left, when the force is placed in the 3.4 area.

If the load is applied to premolar 3.4, the mandibular bone has slightly higher tension values in the loading area. If the load is applied on molar 3.6, a jump in tension values is observed between the dental bridge and abutment tooth (Figure 11).

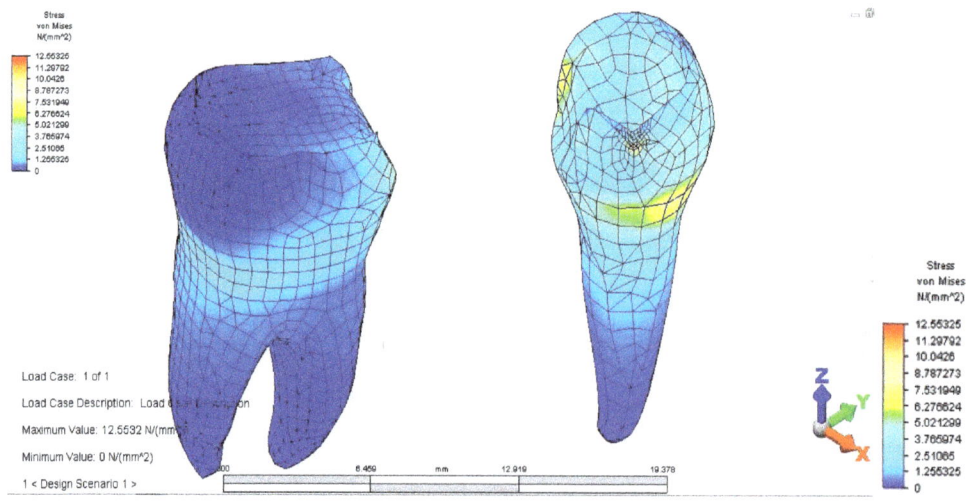

Figure 11. *Cont.*

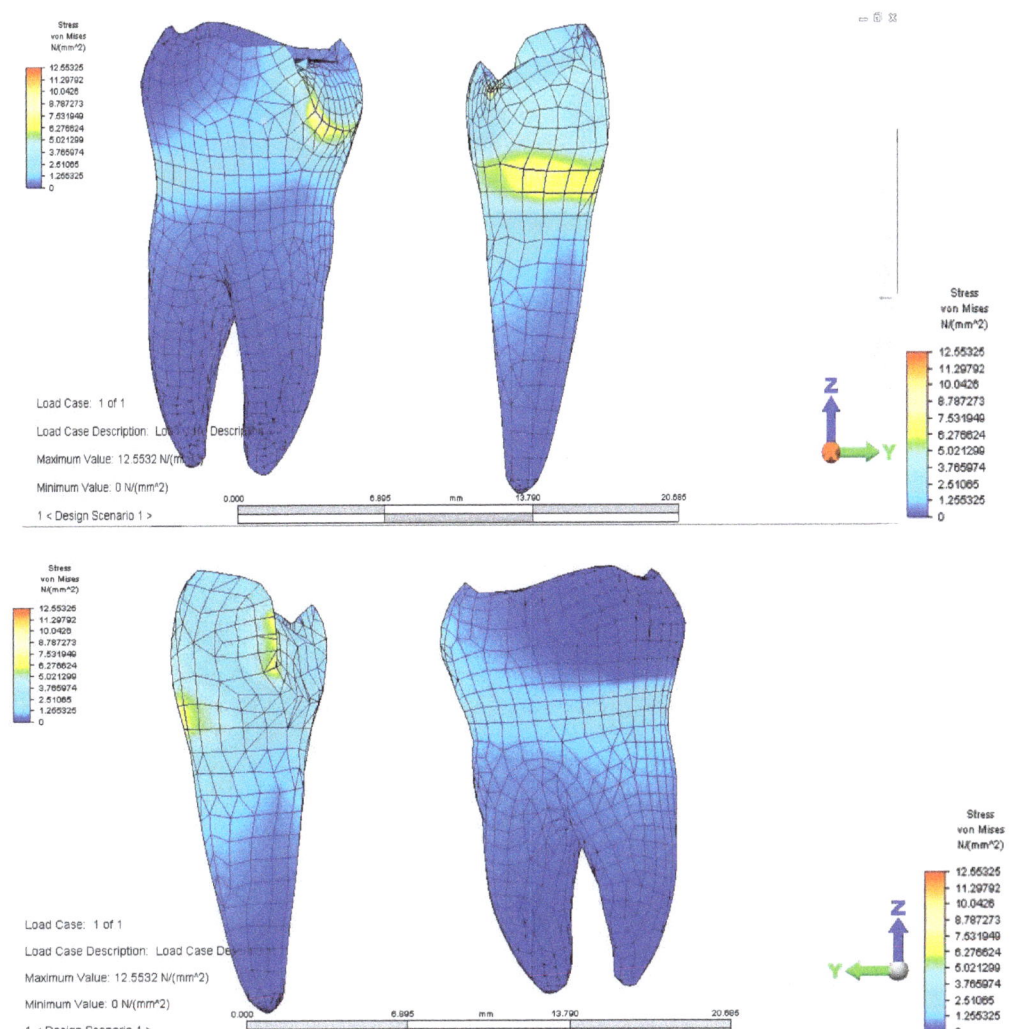

Figure 11. Tension distribution at abutments level in three directions: axonometric, right and left when the force is placed in the 3.4 area.

4. Discussion

The finite element analysis, being a purely mathematical method, has a series of disadvantages because it imposes certain simplifications, which lead to a certain degree of approximation of results. This type of study does not take into account several important parameters that may influence clinical longevity of fixed prosthetic rehabilitation: anisotropy and heterogeneity of alveolar bone, periodontal structures, the viscoelastic response of periodontal structures to functional stresses, coefficients of thermal expansion of analyzed structures, the fatigue effect manifested by prosthetic restoration materials, the complexity of masticatory forces regarding their application point, intensity, summation of action forces with neighboring forces [10]. In addition, the physiological mobility of supporting teeth and particularities of the mandibular bone were not taken into account—density, height, profile [10].

Several authors reported that single implants with fixed prosthetic rehabilitation are the gold standard in the case of mono-edentulism in terms of success rate, patients' reported outcomes and marginal bone loss [11,12].

In our study, the models were subjected to loading with a constant vertical force applied to different points on the occlusal surface of the bridge. However, in a real-life situation, masticatory forces are variable according to an individual's food consistency and muscle activity.

Despite these limitations, finite element analysis provides valuable information and insight into the area of biomechanical research and offers clinical studies a base for development [13,14].

For this study, we chose to model an inlay retained dental bridge due to this being a model less studied in the literature. Furthermore, clinicians are reluctant to use this type of restoration because of a higher clinical failure percentage as a result of being a more technique-sensitive prosthetic solution. On the other hand, it is an important addition to the therapeutic options employed in clinical practice due to the fact that it can be used as a provisional prosthetic rehabilitation method during the 4-months period for bone healing and osteointegration of a dental implant [15]. For this type of prosthesis, the most frequent failure causes are marginal leakage and debonding of restorations when compared to conventional fixed dental prostheses and single crowns supported by implant [16]. Conversely, regarding advantages, it is a more conservative approach, as it can be applied to patients with contraindications of implant placement, and if executed properly, has a high success rate [17].

In our study, when first developing the analysis, we evaluated the distribution of deviatoric stresses (von Mises tensions) at the pontic level, when the reaction force, opposing the muscular force, is located in the missing premolar 3.5 area. This phenomenon occurs due to the difference between the higher modulus of elasticity of the bridge alloy and the modulus of elasticity of abutment teeth. According to Hooke's law, at the same specific deformation, located at the tooth-restoration interface and at different modulus of elasticity, higher stresses result in the prosthetic restoration. We noticed higher tensions developing at the bridge elements junction, the highest value being between teeth 3.6 and 3.5. This high value of stress is explained by the smaller surface of the junction area, which in turn could constitute a potential fracture point. This result is in concordance with other studies in the literature [18].

Moreover, when evaluating the mandible-abutments-dental bridge as a whole, with the application of a reaction force on premolar 3.4, a tension jump is observed between the teeth and the dental bridge. This disparity is more obvious at the premolar 3.5 and the inlay applied on the occlusal surface of the junction with 3.4. The force is further transmitted to the mandibular bone due to increased stress, which upholds a considerably higher tension.

Results similar to ours were obtained in a study that analyzed three interim restorative materials by the FEM method regarding stress resistance. The authors found that the biggest tensile stress magnitude, regardless of the restorative material, was in the region of the prosthetic connector, and the highest stress peak was observed in resin composite, followed by polyetheretherketone and acrylic resin [3].

The asymmetric stress distribution of the dental bridge is due to the placement of the reaction force. Thus when the load is applied to the molar, the value is lower. However, it is higher when the force is placed directly on the pontic. This situation can be explained because the premolar is further from the rotation center and the reaction force, which acts directly on tooth 3.4. Moreover, the abutments have an uneven stress distribution, the value being elevated on the premolar on which the reaction force acts, and an augmented value of tension is noticed in 3.4, towards the coronal area of the root, near the canine. In both analyzed scenarios, there is a tilt of the supporting teeth caused by the direct action of the reaction force on abutments.

A study by Bromicke evaluated the load-bearing capacity, load at initial damage and the failure pattern of posterior resin-bonded fixed dental prostheses to replace a maxillary first molar fabricated from veneered cobalt–chromium, veneered zirconia and monolithic zirconia. Of all tested models, veneered resin-bonded fixed prostheses were more prone to cracking of the veneer component [19]. Other authors observed similar results, who compared fracture resistance of veneered zirconia and metal–ceramic inlay-retained fixed dental prostheses and pinpointed the veneer as being the weakest component [20,21].

Another study, which used the finite element method aimed at testing materials to restore a missing mandibular first molar, found no difference between a posterior inlay-retained full zirconia fixed dental prosthesis and a chromium cobalt substructure, porcelain coating, and adhesive resin as wings. The authors applied a load of 400 N and observed a slight advantage regarding stress-bearing for zirconia, however, not significantly when compared to the chromium cobalt substructure and porcelain coating [22].

A study that analyzed tensile stress between restoration–cement, cement and cement–cavity observed that the tensile stress was directly proportional to the restorative materials elastic modulus. Thus a more rigid cement material increases tensile zones in the layer but decreases the stress between prosthesis and cement. The highest stress concentration between restoration and cement was observed in the molar cavity when compared to the premolar [23].

Other factors that could further influence fracture resistance are the design of the preparation, framework design, and surface treatments of fixed dental prosthesis [24–26]. A study demonstrated that the highest fracture resistance values were observed in the case of the butterfly wings design followed by inlay and box designs [27]. Furthermore, the additional surface treatment by sandblasting and tribochemical silica coating of zirconia surfaces displayed the highest mean fracture resistance values when compared to Er, Cr: YSGG laser [27,28].

Bone density and width of maxillary bones in the edentate area have an important impact on the choice of the future prosthetic appliance, which can be applied. Thus instruments, which analyze these parameters are of utmost relevance, especially when considering implant placement. Furthermore, a mechanical risk evaluation before placing inlay-retained dental bridges or before placing other therapeutic options, especially implant treatment using endoral radiographs, ortopanoramics and cone-beam computer tomography and other paraclinical examinations, should be an obligatory step in treatment planning [29,30].

The most fragile part of the bridge is represented by the junction between the bridge body and aggregation elements; the smaller the section of this area, the more prone it will be to fracture [31,32]. Our study confirmed these results; furthermore, we also emphasized the distribution of forces on the bridge elements and on abutments, highlighting the areas of maximum load, as clinical maneuvers, such as periodontal instrumentation, can further weaken the resistance of abutment teeth [33].

Knowing the vulnerable areas, clinicians will have useful information for designing such a dental bridge and will have the opportunity to increase stability and retention.

The results of our study show that the inlay-retained dental bridge represents an optimal therapeutic solution, in terms of the resistance of abutments, with the added benefit of an important economy of dental tissues.

5. Conclusions

The finite element analysis demonstrates that stress determined by the loading force is not able to cause damage to the prosthetic device or to abutment teeth. However, special attention must be paid to its design, especially in the connection area between the bridge elements, because connectors and retainers represent the weakest parts.

Within the limits of this preliminary study, the inlay-retained dental bridge for single-tooth replacement is a viable alternative, not only from a clinical point of view as the integrity of dental tissues are preserved to a very large extent but also from a biomechanical point of view. Thus, it can be considered an optimal economical solution for treating class III Kennedy edentation in young patients or as a provisional pre-implant rehabilitation option.

Author Contributions: Conceptualization, M.T. and D.D.-P.; methodology, A.V.; software, I.L.; validation, D.D.-P. and A.V.; formal analysis, I.M.; investigation, I.M.; resources, I.L.; data curation, G.A.M.; writing—original draft preparation, D.D.-P.; writing—review and editing, G.A.M.; visualization, G.A.M.; supervision, M.T. and A.V.; project administration, M.T.; funding acquisition, M.T. All authors have read and agreed to the published version of the manuscript.

Funding: This research received no external funding.

Institutional Review Board Statement: Not applicable.

Informed Consent Statement: Not applicable.

Data Availability Statement: The data used to support the findings of this study are available from the corresponding authors upon reasonable request.

Acknowledgments: Ionut Luchian, Ioana Martu have a contribution equal to that of the first author.

Conflicts of Interest: The authors declare no conflict of interest.

References

1. Kawala, M.; Smardz, J.; Adamczyk, L.; Grychowska, N.; Wieckiewicz, M. Selected applications for current polymers in prosthetic dentistry-state of the art. *Curr. Med. Chem.* **2018**, *25*, 6002–6012. [CrossRef]
2. Wieckiewicz, M.; Opitz, V.; Richter, G.; Boening, K.W. Physical properties of polyamide-12 versus PMMA denture base material. *Bio. Med. Res. Int.* **2014**. [CrossRef]
3. Campaner, L.M.; Silveira, M.P.; de Andrade, G.S.; Borges, A.L.; Bottino, M.A.; Dal Piva, A.M.; Lo Giudice, R.; Ausiello, P.; Tribst, J.P. Influence of Polymeric Restorative Materials on the Stress Distribution in Posterior Fixed Partial Dentures: 3D Finite Element Analysis. *Polymers* **2021**, *13*, 758. [CrossRef]
4. Monaco, C.; Cardelli, P.; Bolognesi, M.; Scotti, R.; Ozcan, M. Inlay-retained zirconia fixed dental prosthesis: Clinical and laboratory procedures. *Eur. J. Esthet. Dent.* **2012**, *7*, 48–60.
5. August, D.; Augusti, G.; Borgonovo, A.; Amato, M.; Re, D. Inlay-Retained Fixed Dental Prosthesis: A Clinical Option Using Monolithic Zirconia. *Case. Rep. Dent.* **2014**, *2014*, 1–7. [CrossRef]
6. Rezaei, S.M.; Heidarifar, H.; Arezodar, F.F.; Azary, A.; Mokhtarykhoee, S. Influence of connector width on the stress distribution of posterior bridges under loading. *J. Dent. (Tehran)* **2011**, *8*, 67.
7. Diaconu, D.; Tatarciuc, M.; Vitalariu, A.; Stamatin, O.; Foia, L.; Checherita, L.E. Effect of silver nanoparticles incorporation in dental resins on stress distribution- Finite Element Analysis. *Mat. Plast.* **2014**, *51*, 1571–1574.
8. Bandela, V.; Kanaparthi, S. Finite Element Analysis and Its Applications in Dentistry. In *Finite Element Methods and Their Applications*; IntechOpen: London, UK, 2020.
9. Smielak, B.; Swiniarski, J.; Wolowiec-Korecka, E.; Klimek, L. 2D-Finite element analysis of inlay-, onlay bridges with using various materials. *Arch. Mater. Sci. Eng.* **2016**, *79*, 71–78. [CrossRef]
10. Ferreira, R.C.; Caldas, J.; Paula, G.A.; Albuquerque, R.C.; Almeida, C.M.; Vasconcellos, W.A.; Caldas, R.B. Influence of Surface Area and Geometry of Specimens on Bond Strength in a Microtensile Test: An Analysis by the Three-Dimensional Finite Element Method. *J. Prosthodont.* **2011**, *20*, 456–463. [CrossRef]
11. Sanz-Sánchez, I.; Sanz-Martín, I.; Carrillo de Albornoz, A.; Figuero, E.; Sanz, M. Biological effect of the abutment material on the stability of peri-implant marginal bone levels: A systematic review and meta-analysis. *Clin. Oral. Implants. Res.* **2018**, *29*, 124–144. [CrossRef] [PubMed]
12. Cosola, S.; Marconcini, S.; Boccuzzi, M.; Menchini Fabris, G.B.; Covani, U.; Peñarrocha-Diago, M.; Peñarrocha-Oltra, D. Radiological Outcomes of Bone-Level and Tissue-Level Dental Implants: Systematic Review. *Int. J Environ. Res. Public Health.* **2020**, *17*, 6920. [CrossRef] [PubMed]
13. Kareem, A.A.; Samran, A.; Aswad, M.; Nassani, M.Z. A new design for posterior inlay-retained fixed partial denture. *J. Prosthodont Res.* **2013**, *57*, 146–149.
14. Al-Quran, F.A.; Al-Ghalayini, R.F.; Al-Zu'bi, B.N. Single-tooth replacement: Factors affecting different prosthetic treatment modalities. *BMC Oral health* **2011**, *11*, 1–7. [CrossRef]
15. Cao, J.; Zhou, W.; Shen, S.; Wu, Y.; Wang, X. Implant-supported provisional prosthesis facilitated the minor revision of occlusion and incisor exposure after orthognathic surgery of extended oligodontia in maxilla. *J. Dent. Sci.* **2021**, *16*, 544–548. [CrossRef] [PubMed]
16. Chen, J.; Cai, H.; Suo, L.; Xue, Y.; Wang, J.; Wan, Q. A systematic review of the survival and complication rates of inlay-retained fixed dental prostheses. *J. Dent.* **2017**, *59*, 2–10. [CrossRef]
17. Bömicke, W.; Rathmann, F.; Pilz, M.; Bermejo, J.L.; Waldecker, M.; Ohlmann, B.; Rammelsberg, P.; Zenthöfer, A. Clinical Performance of Posterior Inlay-Retained and Wing-Retained Monolithic Zirconia Resin-Bonded Fixed Partial Dentures: Stage One Results of a Randomized Controlled Trial. *J. Prosthodont* **2020**, 1–10. [CrossRef]
18. Thompson, M.C.; Field, C.J.; Swain, M.V. The all-ceramic, inlay supported fixed partial denture. Part 2. Fixed partial denture design: A finite element analysis. *Aust. Dent. J.* **2011**, *56*, 302–311. [CrossRef]

19. Bömicke, W.; Waldecker, M.; Krisam, J.; Rammelsberg, P.; Rues, S. In vitro comparison of the load-bearing capacity of ceramic and metal-ceramic resin-bonded fixed dental prostheses in the posterior region. *J. Prosthet Dent.* **2018**, *119*, 89–96. [CrossRef]
20. Mohsen, C.A. Fracture Resistance of Three Ceramic Inlay-Retained Fixed Partial Denture Designs. An In Vitro Comparative Study. *J. Prosthodont.* **2010**, *19*, 531–535. [CrossRef]
21. Kılıçarslan, M.A.; Kedici, P.S.; Küçükeşmen, H.C.; Uludağ, B.C. In vitro fracture resistance of posterior metal-ceramic and all-ceramic inlay-retained resin-bonded fixed partial dentures. *J. Prosthet Dent.* **2004**, *92*, 365–370. [CrossRef]
22. Yossef, S.A.; Galal, R.M.; Alqahtani, W.M.; Alluqmani, A.A.; Abdulsamad, M.A.; Alsharabi, O.H.; Smurqandi, E.M. Comparison between two materials for the fabrication of modified design for posterior inlay-retained fixed dental prosthesis: A finite element study. *J. Int. Oral Health* **2018**, *10*, 88.
23. Tribst, J.P.; Dal Piva, A.M.; de Melo, R.M.; Borges, A.L.; Bottino, M.A.; Özcan, M. Influence of restorative material and cement on the stress distribution of posterior resin-bonded fixed dental prostheses: 3D finite element analysis. *J. Mech. Behav. Biomed. Mater.* **2019**, *96*, 279–284. [CrossRef]
24. Bakitian, F.; Papia, E.; Larsson, C.; Vult von Steyern, P. Evaluation of Stress Distribution in Tooth-Supported Fixed Dental Prostheses Made of Translucent Zirconia with Variations in Framework Designs: A Three-Dimensional Finite Element Analysis. *J. Prosthodont.* **2020**, *29*, 315–322. [CrossRef]
25. Malysa, A.; Wezgowiec, J.; Orzeszek, S.; Florjanski, W.; Zietek, M.; Wieckiewicz, M. Effect of Different Surface Treatment Methods on Bond Strength of Dental Ceramics to Dental Hard Tissues: A Systematic Review. *Molecules* **2021**, *26*, 1223. [CrossRef]
26. Ciocan-Pendefunda, A.A.; Martu, M.A.; Antohe, M.E.; Luchian, I.; Martu, I.; Sioustis, I.; Ifteni, G. Indirect composite veneers as a social therapeutic solution. A case report. *Rom. J. Oral. Rehab.* **2018**, *10*, 91–96.
27. Samhan, T.M.; Zaghloul, H. Load to failure of three different monolithic zirconia inlay-retained fixed dental prosthesis designs with three surface treatments. *Braz. Dent. Sci.* **2020**, *23*, 1–10. [CrossRef]
28. Alpízar, M.; Castillo, R.; Chinè, B. Thermal stress analysis by finite elements of a metal-ceramic dental bridge during the cooling phase of a glaze treatment. *J. Mech. Behav. Biomed. Mater.* **2020**, *104*, 103661. [CrossRef]
29. Cosola, S.; Toti, P.; Peñarrocha-Diago, M.; Covani, U.; Brevi, B.C.; Peñarrocha-Oltra, D. Standardization of three-dimensional pose of cylindrical implants from intraoral radiographs: A preliminary study. *BMC Oral Health.* **2021**, *21*, 100. [CrossRef]
30. Kullman, L.; Al-Asfour, A.; Zetterqvist, L.; Andersson, L. Comparison of radiographic bone height assessments in panoramic and intraoral radiographs of implant patients. *Int. J. Oral. Maxillofac. Implants* **2007**, *22*, 96–100.
31. Lakshmi, R.D.; Abraham, A.; Sekar, V.; Hariharan, A. Influence of connector dimensions on the stress distribution of monolithic zirconia and lithium-di-silicate inlay retained fixed dental prostheses—A 3D finite element analysis. *Tanta Dent. J.* **2015**, *12*, 56–64. [CrossRef]
32. Kale, E.; İzgi, A.D.; Niğiz, R. Bond strength evaluation of inlay-retained resin-bonded fixed partial dentures with two different cavity designs and two different adhesive systems: In vitro study. *Balk. J. Dent. Med.* **2020**, *24*, 21–28. [CrossRef]
33. Solomon, S.M.; Timpu, D.; Forna, D.A.; Stefanache, M.A.; Martu, S.; Stoleriu, S. AFM comparative study of root surface morphology after three methods of scaling. *Mater. Plast.* **2016**, *53*, 546–549.